Lecture Notes of the Unione Matematica Italiana

20

More information about this series at http://www.springer.com/series/7172

Editorial Board

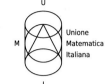

Claudia Bucur • Enrico Valdinoci

Nonlocal Diffusion and Applications

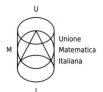

Unione
Matematica
Italiana

Claudia Bucur
Dipartimento di Matematica
 Federigo Enriques
Università degli Studi di Milano
Milano, Italy

Enrico Valdinoci
Dipartimento di Matematica
 Federigo Enriques
Università degli Studi di Milano
Milano, Italy

Consiglio Nazionale delle Ricerche
Istituto di Matematica Applicata e
 Tecnologie Informatiche Enrico Magene
Pavia, Italy

Weierstraß Institut für Angewandte
 Analysis und Stochasitk
Berlin, Germany

University of Melbourne
School of Mathematics and Statistics
Victoria, Australia

ISSN 1862-9113   ISSN 1862-9121   (electronic)
Lecture Notes of the Unione Matematica Italiana
ISBN 978-3-319-28738-6   ISBN 978-3-319-28739-3   (eBook)
DOI 10.1007/978-3-319-28739-3

Library of Congress Control Number: 2016934714

Printed on acid-free paper

This Springer imprint is published by Springer Nature
The registered company is Springer International Publishing AG Switzerland

Preface

The purpose of these pages is to collect a set of notes that are a result of several talks and minicourses delivered here and there in the world (Milan, Cortona, Pisa, Roma, Santiago del Chile, Madrid, Bologna, Porquerolles, and Catania to name a few). We will present here some mathematical models related to nonlocal equations, providing some introductory material and examples.

Of course, these notes and the results presented do not aim to be comprehensive and cannot take into account all the material that would deserve to be included. Even a thorough introduction to nonlocal (or even just fractional) equations goes way beyond the purpose of this book.

Using a metaphor with fine arts, we could say that the picture that we painted here is not even impressionistic, it is just naïf. Nevertheless, we hope that these pages may be of some help to the young researchers of all ages who are willing to have a look at the exciting nonlocal scenario (and who are willing to tolerate the partial and incomplete point of view offered by this modest observation point).

Milano, Italy   Claudia Bucur
Milano, Italy   Enrico Valdinoci
November 2015

Acknowledgments

It is a pleasure to thank Serena Dipierro, Rupert Frank, Richard Mathar, Alexander Nazarov, Joaquim Serra, and Fernando Soria for very interesting and pleasant discussions. We are also indebted with all the participants of the seminars and minicourses from which this set of notes generated for the nice feedback received, and we hope that this work, though somehow sketchy and informal, can be useful to stimulate new discussions and further develop this rich and interesting subject.

Contents

Introduction

In the recent years the fractional Laplace operator has received much attention both in pure and in applied mathematics. Starting from the basics of the nonlocal equations, in this set of notes we will discuss in detail some recent developments in four topics of research on which we focused our attention, namely:

- A problem arising in crystal dislocation (which is related to a classical model introduced by Peierls and Nabarro)
- A problem arising in phase transitions (which is related to a nonlocal version of the classical Allen–Cahn equation)
- The limit interfaces arising in the above nonlocal phase transitions (which turn out to be nonlocal minimal surfaces, as introduced by Caffarelli, Roquejoffre, and Savin)
- A nonlocal version of the Schrödinger equation for standing waves (as introduced by Laskin)

Many fundamental topics slipped completely out of these notes: just to name a few, the topological methods and the fine regularity theory in the fractional cases are not presented here; the fully nonlinear or singular/degenerate equations are not taken into account; only very few applications are discussed briefly; important models such as the quasi-geostrophic equation and the fractional porous media equation are not covered in these notes; we will not consider models arising in game theory such as the nonlocal tug-of-war; the parabolic equations are not taken into account in detail; unique continuation and overdetermined problems will not be studied here, and the link to probability theory that we consider here is not rigorous and only superficial (the reader interested in these important topics may look, for instance, at [8, 10, 14, 15, 17–20, 31, 36–38, 41, 53, 65, 66, 70, 71, 88, 95, 97–100, 109, 113, 127, 133]). Also, a complete discussion of the nonlocal equations in bounded domains is not available here (for this, we refer to the recent survey [119]). In terms of surveys, collections of results, and open problems, we also mention the very nice website [2], which gets[1] constantly updated.

This set of notes is organized as follows. To start with, in Chap. 1, we will give a motivation for the fractional Laplacian (which is the typical nonlocal operator for our framework) that originates from probabilistic considerations. As a matter of fact, no advanced knowledge of probability theory is assumed from the reader, and the topic is dealt with at an elementary level.

In Chap. 2, we will recall some basic properties of the fractional Laplacian, discuss some explicit examples in detail, and point out some structural inequalities that are due to a fractional comparison principle. This part continues with a quite surprising result, which states that every function can be locally approximated by functions with vanishing fractional Laplacian (in sharp contrast with the rigidity of the classical harmonic functions). We also give an example of a function with constant fractional Laplacian on the ball.

In Chap. 3 we deal with extended problems. It is indeed a quite remarkable fact that in many occasions nonlocal operators can be equivalently represented as local (though possibly degenerate or singular) operators in one dimension more. Moreover, as a counterpart, several models arising in a local framework give rise to nonlocal equations, due to boundary effects. So, to introduce the extension problem and give a concrete intuition of it, we will present some models in physics that are naturally set on an extended space to start with and will show their relation with the fractional Laplacian on a trace space. We will also give a detailed justification of this extension procedure by means of the Fourier transform.

As a special example of problems arising in physics that produce a nonlocal equation, we consider a problem related to crystal dislocation, present some mathematical results that have been recently obtained on this topic, and discuss the relation between these results and the observable phenomena.

Chapters 4, 5, and 6 present topics of contemporary research. We will discuss in particular: some phase transition equations of nonlocal type; their limit interfaces, which (below a critical threshold of the fractional parameter) are surfaces that minimize a nonlocal perimeter functional; and some nonlocal equations arising in quantum mechanics.

We remark that the introductory part of these notes is not intended to be separated from the one which is more research oriented, namely, even the chapters whose main goal is to develop the basics of the theory contain some parts related to contemporary research trends.

---

[1]It seems to be known that Plato did not like books because they cannot respond to questions. He might have liked websites.

Chapter 1
A Probabilistic Motivation

The fractional Laplacian will be the main operator studied in this book. We consider a function $u: \mathbb{R}^n \to \mathbb{R}$ (which is supposed[1] to be regular enough) and a fractional parameter $s \in (0, 1)$. Then, the fractional Laplacian of u is given by

$$(-\Delta)^s u(x) = \frac{C(n,s)}{2} \int_{\mathbb{R}^n} \frac{2u(x) - u(x+y) - u(x-y)}{|y|^{n+2s}} \, dy, \qquad (1.1)$$

where $C(n,s)$ is a dimensional[2] constant.

One sees from (1.1) that $(-\Delta)^s$ is an operator of order $2s$, namely, it arises from a differential quotient of order $2s$ weighted in the whole space. Different fractional operators have been considered in literature (see e.g. [39, 111, 128]), and all of them come from interesting problems in pure or/and applied mathematics. We will focus here on the operator in (1.1) and we will motivate it by probabilistic considerations (as a matter of fact, many other motivations are possible).

The probabilistic model under consideration is a random process that allows long jumps (in further generality, it is known that the fractional Laplacian is an infinitesimal generator of Lèvy processes, see e.g. [7, 13] for further details). A more detailed mathematical introduction to the fractional Laplacian is then presented in the subsequent Sect. 2.1.

---

[1] To write (1.1) it is sufficient, for simplicity, to take here u in the Schwartz space $\mathscr{S}(\mathbb{R}^n)$ of smooth and rapidly decaying functions, or in $C^2(\mathbb{R}^n) \cap L^\infty(\mathbb{R}^n)$. We refer to [131] for a refinement of the space of definition.

[2] The explicit value of $C(n,s)$ is usually unimportant. Nevertheless, we will compute its value explicitly in formulas (2.10) and (2.15). The reason for which it is convenient to divide $C(n,s)$ by a factor 2 in (1.1) will be clear later on, in formula (2.5).

© Springer International Publishing Switzerland 2016
C. Bucur, E. Valdinoci, *Nonlocal Diffusion and Applications*, Lecture Notes of the Unione Matematica Italiana 20, DOI 10.1007/978-3-319-28739-3_1

1.1   The Random Walk with Arbitrarily Long Jumps

We will show here that the fractional heat equation (i.e. the "typical" equation that drives the fractional diffusion and that can be written, up to dimensional constants, as $\partial_t u + (-\Delta)^s u = 0$) naturally arises from a probabilistic process in which a particle moves randomly in the space subject to a probability that allows long jumps with a polynomial tail.

For this scope, we introduce a probability distribution on the natural numbers $\mathbb{N}^* := \{1, 2, 3, \cdots\}$ as follows. If $I \subseteq \mathbb{N}^*$, then the probability of I is defined to be

$$P(I) := c_s \sum_{k \in I} \frac{1}{|k|^{1+2s}}.$$

The constant c_s is taken in order to normalize P to be a probability measure. Namely, we take

$$c_s := \left(\sum_{k \in \mathbb{N}^*} \frac{1}{|k|^{1+2s}} \right)^{-1},$$

so that we have $P(\mathbb{N}^*) = 1$.

Now we consider a particle that moves in $\mathbb{R}^n$ according to a probabilistic process. The process will be discrete both in time and space (in the end, we will formally take the limit when these time and space steps are small). We denote by τ the discrete time step, and by h the discrete space step. We will take the scaling $\tau = h^{2s}$ and we denote by $u(x, t)$ the probability of finding the particle at the point x at time t.

The particle in $\mathbb{R}^n$ is supposed to move according to the following probabilistic law: at each time step τ, the particle selects randomly both a direction $v \in \partial B_1$, according to the uniform distribution on ∂B_1, and a natural number $k \in \mathbb{N}^*$, according to the probability law P, and it moves by a discrete space step khv. Notice that long jumps are allowed with small probability. Then, if the particle is at time t at the point x_0 and, following the probability law, it picks up a direction $v \in \partial B_1$ and a natural number $k \in \mathbb{N}^*$, then the particle at time $t + \tau$ will lie at $x_0 + khv$.

Now, the probability $u(x, t + \tau)$ of finding the particle at x at time $t + \tau$ is the sum of the probabilities of finding the particle somewhere else, say at $x + khv$, for some direction $v \in \partial B_1$ and some natural number $k \in \mathbb{N}^*$, times the probability of having selected such a direction and such a natural number. This translates into

$$u(x, t + \tau) = \frac{c_s}{|\partial B_1|} \sum_{k \in \mathbb{N}^*} \int_{\partial B_1} \frac{u(x + khv, t)}{|k|^{1+2s}} \, d\mathcal{H}^{n-1}(v).$$

Notice that the factor $c_s/|\partial B_1|$ is a normalizing probability constant, hence we subtract $u(x,t)$ and we obtain

$$u(x, t+\tau) - u(x,t) = \frac{c_s}{|\partial B_1|} \sum_{k \in \mathbb{N}^*} \int_{\partial B_1} \frac{u(x+khv, t)}{|k|^{1+2s}} \, d\mathcal{H}^{n-1}(v) - u(x,t)$$

$$= \frac{c_s}{|\partial B_1|} \sum_{k \in \mathbb{N}^*} \int_{\partial B_1} \frac{u(x+khv, t) - u(x,t)}{|k|^{1+2s}} \, d\mathcal{H}^{n-1}(v).$$

As a matter of fact, by symmetry, we can change v to $-v$ in the integral above, so we find that

$$u(x, t+\tau) - u(x,t) = \frac{c_s}{|\partial B_1|} \sum_{k \in \mathbb{N}^*} \int_{\partial B_1} \frac{u(x-khv, t) - u(x,t)}{|k|^{1+2s}} \, d\mathcal{H}^{n-1}(v).$$

Then we can sum up these two expressions (and divide by 2) and obtain that

$$u(x, t+\tau) - u(x,t)$$
$$= \frac{c_s}{2\,|\partial B_1|} \sum_{k \in \mathbb{N}^*} \int_{\partial B_1} \frac{u(x+khv, t) + u(x-khv, t) - 2u(x,t)}{|k|^{1+2s}} \, d\mathcal{H}^{n-1}(v).$$

Now we divide by $\tau = h^{2s}$, we recognize a Riemann sum, we take a formal limit and we use polar coordinates, thus obtaining:

$$\partial_t u(x,t) \simeq \frac{u(x, t+\tau) - u(x,t)}{\tau}$$

$$= \frac{c_s h}{2\,|\partial B_1|} \sum_{k \in \mathbb{N}^*} \int_{\partial B_1} \frac{u(x+khv, t) + u(x-khv, t) - 2u(x,t)}{|hk|^{1+2s}} d\mathcal{H}^{n-1}(v)$$

$$\simeq \frac{c_s}{2\,|\partial B_1|} \int_0^{+\infty} \int_{\partial B_1} \frac{u(x+rv, t) + u(x-rv, t) - 2u(x,t)}{|r|^{1+2s}} d\mathcal{H}^{n-1}(v)\, dr$$

$$= \frac{c_s}{2\,|\partial B_1|} \int_{\mathbb{R}^n} \frac{u(x+y, t) + u(x-y, t) - 2u(x,t)}{|y|^{n+2s}} \, dy$$

$$= -c_{n,s}\, (-\Delta)^s u(x,t)$$

for a suitable $c_{n,s} > 0$.

This shows that, at least formally, for small time and space steps, the above probabilistic process approaches a fractional heat equation.

We observe that processes of this type occur in nature quite often, see in particular the biological observations in [90, 140], other interesting observations in [118, 126, 142] and the mathematical discussions in [84, 93, 104, 107, 110].

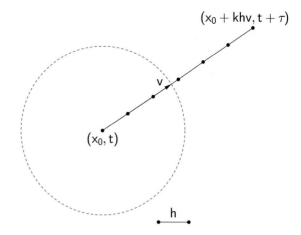

Fig. 1.1 The random walk with jumps

Roughly speaking, let us say that it is not unreasonable that a predator may decide to use a nonlocal dispersive strategy to hunt its preys more efficiently (or, equivalently, that the natural selection may favor some kind of nonlocal diffusion): small fishes will not wait to be eaten by a big fish once they have seen it, so it may be more convenient for the big fish just to pick up a random direction, move rapidly in that direction, stop quickly and eat the small fishes there (if any) and then go on with the hunt. And this "hit-and-run" hunting procedure seems quite related to that described in Fig. 1.1.

1.2   A Payoff Model

Another probabilistic motivation for the fractional Laplacian arises from a payoff approach. Suppose to move in a domain Ω according to a random walk with jumps as discussed in Sect. 1.1. Suppose also that exiting the domain Ω for the first time by jumping to an outside point $y \in \mathbb{R}^n \setminus \Omega$, means earning $u_0(y)$ sestertii. A relevant question is, of course, how rich we expect to become in this way. That is, if we start at a given point $x \in \Omega$ and we denote by $u(x)$ the amount of sestertii that we expect to gain, is there a way to obtain information on u?

The answer is that (in the right scale limit of the random walk with jumps presented in Sect. 1.1) the expected payoff u is determined by the equation

$$\begin{cases} (-\Delta)^s u = 0 & \text{in } \Omega, \\ u = u_0 & \text{in } \mathbb{R}^n \setminus \Omega. \end{cases} \tag{1.2}$$

To better explain this, let us fix a point $x \in \Omega$. The expected value of the payoff at x is the average of all the payoffs at the points $\tilde{x}$ from which one can reach x, weighted by the probability of the jumps. That is, by writing $\tilde{x} = x + khv$, with $v \in \partial B_1$,

$k \in \mathbb{N}^*$ and $h > 0$, as in the previous Sect. 1.1, we have that the probability of jump is $\dfrac{c_s}{|\partial B_1|\,|k|^{1+2s}}$. This leads to the formula

$$u(x) = \frac{c_s}{|\partial B_1|} \sum_{k\in\mathbb{N}^*} \int_{\partial B_1} \frac{u(x + khv)}{|k|^{1+2s}}\, d\mathcal{H}^{n-1}(v).$$

By changing v into $-v$ we obtain

$$u(x) = \frac{c_s}{|\partial B_1|} \sum_{k\in\mathbb{N}^*} \int_{\partial B_1} \frac{u(x - khv)}{|k|^{1+2s}}\, d\mathcal{H}^{n-1}(v)$$

and so, by summing up,

$$2u(x) = \frac{c_s}{|\partial B_1|} \sum_{k\in\mathbb{N}^*} \int_{\partial B_1} \frac{u(x + khv) + u(x - khv)}{|k|^{1+2s}}\, d\mathcal{H}^{n-1}(v).$$

Since the total probability is 1, we can subtract $2u(x)$ to both sides and obtain that

$$0 = \frac{c_s}{|\partial B_1|} \sum_{k\in\mathbb{N}^*} \int_{\partial B_1} \frac{u(x + khv) + u(x - khv) - 2u(x)}{|k|^{1+2s}}\, d\mathcal{H}^{n-1}(v).$$

We can now divide by h^{1+2s} and recognize a Riemann sum, which, after passing to the limit as $h \searrow 0$, gives $0 = -(-\Delta)^s u(x)$, that is (1.2).

Chapter 2
An Introduction to the Fractional Laplacian

We introduce here some preliminary notions on the fractional Laplacian and on fractional Sobolev spaces. Moreover, we present an explicit example of an s-harmonic function on the positive half-line $\mathbb{R}_+$, an example of a function with constant Laplacian on the ball, discuss some maximum principles and a Harnack inequality, and present a quite surprising local density property of s-harmonic functions into the space of smooth functions.

2.1 Preliminary Notions

We introduce here the fractional Laplace operator, the fractional Sobolev spaces and give some useful pieces of notation. We also refer to [57] for further details related to the topic.

We consider the Schwartz space of rapidly decaying functions defined as

$$\mathscr{S}(\mathbb{R}^n) := \left\{ f \in C^\infty(\mathbb{R}^n) \ \middle|\ \forall \alpha, \ \beta \in \mathbf{N}_0^n, \ \sup_{x \in \mathbb{R}^n} |x^\alpha \partial_\beta f(x)| < \infty \right\}.$$

For any $f \in \mathscr{S}(\mathbb{R}^n)$, denoting the space variable $x \in \mathbb{R}^n$ and the frequency variable $\xi \in \mathbb{R}^n$, the Fourier transform and the inverse Fourier transform are defined, respectively, as

$$\hat{f}(\xi) := \mathscr{F}f(\xi) := \int_{\mathbb{R}^n} f(x)e^{-2\pi i x \cdot \xi} \, dx \tag{2.1}$$

and

$$f(x) = \mathscr{F}^{-1}\hat{f}(x) = \int_{\mathbb{R}^n} \hat{f}(\xi)e^{2\pi i x \cdot \xi} \, d\xi. \tag{2.2}$$

© Springer International Publishing Switzerland 2016
C. Bucur, E. Valdinoci, *Nonlocal Diffusion and Applications*, Lecture Notes of the Unione Matematica Italiana 20, DOI 10.1007/978-3-319-28739-3_2

Another useful notion is the one of principal value, namely we consider the definition

$$\text{P.V.} \int_{\mathbb{R}^n} \frac{u(x) - u(y)}{|x - y|^{n+2s}} \, dy := \lim_{\varepsilon \to 0} \int_{\mathbb{R}^n \setminus B_\varepsilon(x)} \frac{u(x) - u(y)}{|x - y|^{n+2s}} \, dy. \tag{2.3}$$

Notice indeed that the integrand above is singular when y is in a neighborhood of x, and this singularity is, in general, not integrable (in the sense of Lebesgue): indeed notice that, near x, we have that $u(x) - u(y)$ behaves at the first order like $\nabla u(x) \cdot (x - y)$, hence the integral above behaves at the first order like

$$\frac{\nabla u(x) \cdot (x - y)}{|x - y|^{n+2s}} \tag{2.4}$$

whose absolute value gives an infinite integral near x (unless either $\nabla u(x) = 0$ or $s < 1/2$).

The idea of the definition in (2.3) is that the term in (2.4) averages out in a neighborhood of x by symmetry, since the term is odd with respect to x, and so it does not contribute to the integral if we perform it in a symmetric way. In a sense, the principal value in (2.3) kills the first order of the function at the numerator, which produces a linear growth, and focuses on the second order remainders.

The notation in (2.3) allows us to write (1.1) in the following more compact form:

$$
\begin{aligned}
(-\Delta)^s u(x) &= \frac{C(n,s)}{2} \int_{\mathbb{R}^n} \frac{2u(x) - u(x+y) - u(x-y)}{|y|^{n+2s}} \, dy \\
&= \frac{C(n,s)}{2} \lim_{\varepsilon \to 0} \int_{\mathbb{R}^n \setminus B_\varepsilon} \frac{2u(x) - u(x+y) - u(x-y)}{|y|^{n+2s}} \, dy \\
&= \frac{C(n,s)}{2} \lim_{\varepsilon \to 0} \left[\int_{\mathbb{R}^n \setminus B_\varepsilon} \frac{u(x) - u(x+y)}{|y|^{n+2s}} \, dy + \int_{\mathbb{R}^n \setminus B_\varepsilon} \frac{u(x) - u(x-y)}{|y|^{n+2s}} \, dy \right] \\
&= \frac{C(n,s)}{2} \lim_{\varepsilon \to 0} \left[\int_{\mathbb{R}^n \setminus B_\varepsilon(x)} \frac{u(x) - u(\eta)}{|x - \eta|^{n+2s}} \, d\eta + \int_{\mathbb{R}^n \setminus B_\varepsilon(x)} \frac{u(x) - u(\zeta)}{|x - \zeta|^{n+2s}} \, d\zeta \right] \\
&= C(n,s) \lim_{\varepsilon \to 0} \int_{\mathbb{R}^n \setminus B_\varepsilon(x)} \frac{u(x) - u(\eta)}{|x - \eta|^{n+2s}} \, d\eta,
\end{aligned}
$$

where the changes of variable $\eta := x + y$ and $\zeta := x - y$ were used, i.e.

$$(-\Delta)^s u(x) = C(n,s) \, \text{P.V.} \int_{\mathbb{R}^n} \frac{u(x) - u(y)}{|x - y|^{n+2s}} \, dy. \tag{2.5}$$

The simplification above also explains why it was convenient to write (1.1) with the factor 2 dividing $C(n,s)$. Notice that the expression in (1.1) does not require the P.V. formulation since, for instance, taking $u \in L^\infty(\mathbb{R}^n)$ and locally C^2, using a Taylor expansion of u in B_1, one observes that

$$\int_{\mathbb{R}^n} \frac{|2u(x) - u(x+y) - u(x-y)|}{|y|^{n+2s}} \, dy$$

$$\leq \|u\|_{L^\infty(\mathbb{R}^n)} \int_{\mathbb{R}^n \setminus B_1} |y|^{-n-2s} \, dy + \int_{B_1} \frac{|D^2 u(x)||y|^2}{|y|^{n+2s}} \, dy$$

$$\leq \|u\|_{L^\infty(\mathbb{R}^n)} \int_{\mathbb{R}^n \setminus B_1} |y|^{-n-2s} \, dy + \|D^2 u\|_{L^\infty(\mathbb{R}^n)} \int_{B_1} |y|^{-n-2s+2} \, dy,$$

and the integrals above provide a finite quantity.

Formula (2.5) has also a stimulating analogy with the classical Laplacian. Namely, the classical Laplacian (up to normalizing constants) is the measure of the infinitesimal displacement of a function in average (this is the "elastic" property of harmonic functions, whose value at a given point tends to revert to the average in a ball). Indeed, by canceling the odd contributions, and using that

$$\int_{B_r(x)} |x - y|^2 \, dy = \sum_{k=1}^n \int_{B_r(x)} (x_k - y_k)^2 \, dy = n \int_{B_r(x)} (x_i - y_i)^2 \, dy,$$

$$\text{for any } i \in \{1, \dots, n\},$$

we see that

$$\lim_{r \to 0} \frac{1}{r^2} \left(u(x) - \frac{1}{|B_r(x)|} \int_{B_r(x)} u(y) \, dy \right)$$

$$= \lim_{r \to 0} -\frac{1}{r^2 |B_r(x)|} \int_{B_r(x)} (u(y) - u(x)) \, dy$$

$$= \lim_{r \to 0} -\frac{1}{r^{n+2} |B_1|} \int_{B_r(x)} \nabla u(x) \cdot (x - y) + \frac{1}{2} D^2 u(x)(x - y) \cdot (x - y)$$

$$\qquad + \mathcal{O}(|x - y|^3) \, dy$$

$$= \lim_{r \to 0} -\frac{1}{2r^{n+2} |B_1|} \sum_{i,j=1}^n \int_{B_r(x)} \partial^2_{i,j} u(x) \, (x_i - y_i)(x_j - y_j) \, dy \qquad (2.6)$$

$$= \lim_{r \to 0} -\frac{1}{2r^{n+2} |B_1|} \sum_{i=1}^n \int_{B_r(x)} \partial^2_{i,i} u(x) \, (x_i - y_i)^2 \, dy$$

$$= \lim_{r \to 0} -\frac{1}{2n \, r^{n+2} |B_1|} \sum_{i=1}^n \partial^2_{i,i} u(x) \int_{B_r(x)} |x - y|^2 \, dy$$

$$= -C_n \Delta u(x),$$

for some $C_n > 0$. In this spirit, when we compare the latter formula with (2.5), we can think that the fractional Laplacian corresponds to a weighted average of the function's oscillation. While the average in (2.6) is localized in the vicinity of a point x, the one in (2.5) occurs in the whole space (though it decays at infinity). Also, the spacial homogeneity of the average in (2.6) has an extra factor that is proportional to the space variables to the power -2, while the corresponding power in the average in (2.5) is $-2s$ (and this is consistent for $s \to 1$).

Furthermore, for $u \in \mathscr{S}(\mathbb{R}^n)$ the fractional Laplace operator can be expressed in Fourier frequency variables multiplied by $(2\pi|\xi|)^{2s}$, as stated in the following lemma.

Lemma 2.1 *We have that*

$$(-\Delta)^s u(x) = \mathscr{F}^{-1}\big((2\pi|\xi|)^{2s}\hat{u}(\xi)\big). \tag{2.7}$$

Roughly speaking, formula (2.7) characterizes the fractional Laplace operator in the Fourier space, by taking the s-power of the multiplier associated to the classical Laplacian operator. Indeed, by using the inverse Fourier transform, one has that

$$-\Delta u(x) = -\Delta(\mathscr{F}^{-1}(\hat{u}))(x) = -\Delta \int_{\mathbb{R}^n} \hat{u}(\xi)e^{2\pi ix\cdot\xi}\,d\xi$$

$$= \int_{\mathbb{R}^n} (2\pi|\xi|)^2\hat{u}(\xi)e^{2\pi ix\cdot\xi}\,d\xi = \mathscr{F}^{-1}\big((2\pi|\xi|)^2\hat{u}(\xi)\big),$$

which gives that the classical Laplacian acts in a Fourier space as a multiplier of $(2\pi|\xi|)^2$. From this and Lemma 2.1 it also follows that the classical Laplacian is the limit case of the fractional one, namely for any $u \in \mathscr{S}(\mathbb{R}^n)$

$$\lim_{s\to 1}(-\Delta)^s u = -\Delta u \quad \text{and also} \quad \lim_{s\to 0}(-\Delta)^s u = u.$$

Let us now prove that indeed the two formulations (1.1) and (2.7) are equivalent.

Proof (Proof of Lemma 2.1) Consider identity (1.1) and apply the Fourier transform to obtain

$$\mathscr{F}\big((-\Delta)^s u(x)\big) = \frac{C(n,s)}{2} \int_{\mathbb{R}^n} \frac{\mathscr{F}\Big(2u(x) - u(x+y) - u(x-y)\Big)}{|y|^{n+2s}}\,dy$$

$$= \frac{C(n,s)}{2} \int_{\mathbb{R}^n} \hat{u}(\xi)\frac{2 - e^{2\pi i\xi\cdot y} - e^{-2\pi i\xi\cdot y}}{|y|^{n+2s}}\,dy \tag{2.8}$$

$$= C(n,s)\,\hat{u}(\xi) \int_{\mathbb{R}^n} \frac{1 - \cos(2\pi\xi\cdot y)}{|y|^{n+2s}}\,dy.$$

Now, we use the change of variable $z = |\xi|y$ and obtain that

$$J(\xi) := \int_{\mathbb{R}^n} \frac{1 - \cos(2\pi\xi \cdot y)}{|y|^{n+2s}} \, dy$$

$$= |\xi|^{2s} \int_{\mathbb{R}^n} \frac{1 - \cos \frac{2\pi\xi}{|\xi|} \cdot z}{|z|^{n+2s}} \, dz.$$

Now we use that J is rotationally invariant. More precisely, we consider a rotation R that sends $e_1 = (1, 0, \ldots, 0)$ into $\xi/|\xi|$, that is $Re_1 = \xi/|\xi|$, and we call R^T its transpose. Then, by using the change of variables $\omega = R^T z$ we have that

$$J(\xi) = |\xi|^{2s} \int_{\mathbb{R}^n} \frac{1 - \cos(2\pi Re_1 \cdot z)}{|z|^{n+2s}} \, dz$$

$$= |\xi|^{2s} \int_{\mathbb{R}^n} \frac{1 - \cos(2\pi R^T z \cdot e_1)}{|R^T z|^{n+2s}} \, dz$$

$$= |\xi|^{2s} \int_{\mathbb{R}^n} \frac{1 - \cos(2\pi\omega_1)}{|\omega|^{n+2s}} \, d\omega.$$

Changing variables $\tilde{\omega} = 2\pi\omega$ (we still write ω as a variable of integration), we obtain that

$$J(\xi) = (2\pi|\xi|)^{2s} \int_{\mathbb{R}^n} \frac{1 - \cos\omega_1}{|\omega|^{n+2s}} \, d\omega. \tag{2.9}$$

Notice that this latter integral is finite. Indeed, integrating outside the ball B_1 we have that

$$\int_{\mathbb{R}^n \setminus B_1} \frac{|1 - \cos\omega_1|}{|\omega|^{n+2s}} \, d\omega \le \int_{\mathbb{R}^n \setminus B_1} \frac{2}{|\omega|^{n+2s}} < \infty,$$

while inside the ball we can use the Taylor expansion of the cosine function and observe that

$$\int_{B_1} \frac{|1 - \cos\omega_1|}{|\omega|^{n+2s}} \, d\omega \le \int_{B_1} \frac{|\omega|^2}{|\omega|^{n+2s}} \, d\omega \le \int_{B_1} \frac{d\omega}{|\omega|^{n+2s-2}} < \infty.$$

Therefore, by taking

$$C(n, s) := \left(\int_{\mathbb{R}^n} \frac{1 - \cos\omega_1}{|\omega|^{n+2s}} \, d\omega \right)^{-1} \tag{2.10}$$

it follows from (2.9) that

$$J(\xi) = \frac{(2\pi|\xi|)^{2s}}{C(n, s)}.$$

By inserting this into (2.8), we obtain that

$$\mathscr{F}\left((-\Delta)^s u(x)\right) = C(n,s)\,\hat{u}(\xi)\,J(\xi) = (2\pi|\xi|)^{2s}\hat{u}(\xi),$$

which concludes the proof.

Notice that the renormalization constant $C(n,s)$ introduced in (1.1) is now computed in (2.10).

Another approach to the fractional Laplacian comes from the theory of semigroups (or, equivalently from the fractional calculus arising in subordination identities). This technique is classical (see [143]), but it has also been efficiently used in recent research papers (see for instance [33, 51, 134]). Roughly speaking, the main idea underneath the semigroup approach comes from the following explicit formulas for the Euler's function: for any $\lambda > 0$, one uses an integration by parts and the substitution $\tau = \lambda t$ to see that

$$-s\Gamma(-s) = \Gamma(1-s)$$

$$= \int_0^{+\infty} \tau^{-s} e^{-\tau}\, d\tau$$

$$= -\int_0^{+\infty} \tau^{-s} \frac{d}{d\tau}(e^{-\tau} - 1)\, d\tau$$

$$= -s\int_0^{+\infty} \tau^{-s-1}(e^{-\tau} - 1)\, d\tau$$

$$= -s\lambda^{-s}\int_0^{+\infty} t^{-s-1}(e^{-\lambda t} - 1)\, dt,$$

that is

$$\lambda^s = \frac{1}{\Gamma(-s)}\int_0^{+\infty} t^{-s-1}(e^{-\lambda t} - 1)\, dt. \qquad (2.11)$$

Then one applies formally this identity to $\lambda := -\Delta$. Of course, this formal step needs to be justified, but if things go well one obtains

$$(-\Delta)^s = \frac{1}{\Gamma(-s)}\int_0^{+\infty} t^{-s-1}(e^{t\Delta} - 1)\, dt,$$

that is (interpreting 1 as the identity operator)

$$(-\Delta)^s u(x) = \frac{1}{\Gamma(-s)}\int_0^{+\infty} t^{-s-1}(e^{t\Delta}u(x) - u(x))\, dt. \qquad (2.12)$$

Formally, if $U(x,t) := e^{t\Delta}u(x)$, we have that $U(x,0) = u(x)$ and

$$\partial_t U = \frac{\partial}{\partial t}(e^{t\Delta}u(x)) = \Delta e^{t\Delta}u(x) = \Delta U,$$

that is $U(x,t) = e^{t\Delta}u(x)$ can be interpreted as the solution of the heat equation with initial datum u. We indeed point out that these formal computations can be justified:

Lemma 2.2 *Formula (2.12) holds true. That is, if $u \in \mathscr{S}(\mathbb{R}^n)$ and $U = U(x,t)$ is the solution of the heat equation*

$$\begin{cases} \partial_t U = \Delta U & \text{in } t > 0, \\ U\big|_{t=0} = u, \end{cases}$$

then

$$(-\Delta)^s u(x) = \frac{1}{\Gamma(-s)} \int_0^{+\infty} t^{-s-1}(U(x,t) - u(x))\, dt. \tag{2.13}$$

Proof From Theorem 1 on page 47 in [69] we know that U is obtained by Gaussian convolution with unit mass, i.e.

$$U(x,t) = \int_{\mathbb{R}^n} G(x-y,t)\, u(y)\, dy = \int_{\mathbb{R}^n} G(y,t)\, u(x-y)\, dy, \tag{2.14}$$

$$\text{where } G(x,t) := (4\pi t)^{-n/2} e^{-|x|^2/(4t)}.$$

As a consequence, using the substitution $\tau := |y|^2/(4t)$,

$$\int_0^{+\infty} t^{-s-1}(U(x,t) - u(x))\, dt$$

$$= \int_0^{+\infty} \left[\int_{\mathbb{R}^n} t^{-s-1} G(y,t) \left(u(x-y) - u(x) \right) dy \right] dt$$

$$= \int_0^{+\infty} \left[\int_{\mathbb{R}^n} (4\pi t)^{-n/2} t^{-s-1} e^{-|y|^2/(4t)} \left(u(x-y) - u(x) \right) dy \right] dt$$

$$= \int_0^{+\infty} \left[\int_{\mathbb{R}^n} \tau^{n/2}(\pi|y|^2)^{-n/2}|y|^{-2s}(4\tau)^{s+1} e^{-\tau} \left(u(x-y) - u(x) \right) dy \right] \frac{d\tau}{4\tau^2}$$

$$= 2^{2s-1}\pi^{-n/2} \int_0^{+\infty} \left[\int_{\mathbb{R}^n} \tau^{\frac{n}{2}+s-1} e^{-\tau} \frac{u(x+y) + u(x-y) - 2u(x)}{|y|^{n+2s}}\, dy \right] d\tau.$$

Now we notice that

$$\int_0^{+\infty} \tau^{\frac{n}{2}+s-1} e^{-\tau}\, d\tau = \Gamma\left(\frac{n}{2} + s\right),$$

so we obtain that

$$\int_0^{+\infty} t^{-s-1}(U(x,t) - u(x))\, dt$$

$$= 2^{2s-1}\pi^{-n/2}\Gamma\left(\frac{n}{2}+s\right)\int_{\mathbb{R}^n}\frac{u(x+y)+u(x-y)-2u(x)}{|y|^{n+2s}}\,dy$$

$$= -\frac{2^{2s}\,\pi^{-n/2}\,\Gamma\left(\frac{n}{2}+s\right)}{C(n,s)}(-\Delta)^s u(x).$$

This proves (2.13), by choosing $C(n,s)$ appropriately. And, as a matter of fact, gives the explicit value of the constant $C(n,s)$ as

$$C(n,s) = -\frac{2^{2s}\,\Gamma\left(\frac{n}{2}+s\right)}{\pi^{n/2}\Gamma(-s)} = \frac{2^{2s}\,s\,\Gamma\left(\frac{n}{2}+s\right)}{\pi^{n/2}\Gamma(1-s)}, \tag{2.15}$$

where we have used again that $\Gamma(1-s) = -s\Gamma(-s)$, for any $s \in (0,1)$.

It is worth pointing out that the renormalization constant $C(n,s)$ introduced in (1.1) has now been explicitly computed in (2.15). Notice that the choices of $C(n,s)$ in (2.10) and (2.15) must agree (since we have computed the fractional Laplacian in two different ways): for a computation that shows that the quantity in (2.10) coincides with the one in (2.15), see Theorem 3.9 in [22]. For completeness, we give below a direct proof that the settings in (2.10) and (2.15) are the same, by using Fourier methods and (2.11).

Lemma 2.3 *For any $n \in \mathbb{N}$, $n \geq 1$, and $s \in (0,1)$, we have that*

$$\int_{\mathbb{R}^n}\frac{1-\cos(2\pi\omega_1)}{|\omega|^{n+2s}}\,d\omega = \frac{\pi^{\frac{n}{2}+2s}\,\Gamma(1-s)}{s\,\Gamma\left(\frac{n}{2}+s\right)}. \tag{2.16}$$

Equivalently, we have that

$$\int_{\mathbb{R}^n}\frac{1-\cos\omega_1}{|\omega|^{n+2s}}\,d\omega = \frac{\pi^{\frac{n}{2}}\,\Gamma(1-s)}{2^{2s}s\,\Gamma\left(\frac{n}{2}+s\right)}. \tag{2.17}$$

Proof Of course, formula (2.16) is equivalent to (2.17) (after the substitution $\tilde{\omega} := 2\pi\omega$). Strictly speaking, in Lemma 2.1 (compare (1.1), (2.7), and (2.10)) we have proved that

$$\frac{1}{2\int_{\mathbb{R}^n}\frac{1-\cos\omega_1}{|\omega|^{n+2s}}\,d\omega}\int_{\mathbb{R}^n}\frac{2u(x)-u(x+y)-u(x-y)}{|y|^{n+2s}}\,dy = \mathscr{F}^{-1}\left((2\pi|\xi|)^{2s}\hat{u}(\xi)\right). \tag{2.18}$$

Similarly, by means of Lemma 2.2 (compare (1.1), (2.13) and (2.15)) we know that

$$\frac{2^{2s-1}\, s\, \Gamma\left(\frac{n}{2}+s\right)}{\pi^{n/2}\Gamma(1-s)} \int_{\mathbb{R}^n} \frac{2u(x)-u(x+y)-u(x-y)}{|y|^{n+2s}}\, dy$$

$$= \frac{1}{\Gamma(-s)} \int_0^{+\infty} t^{-s-1}\left(U(x,t)-u(x)\right) dt.$$

(2.19)

Moreover (see (2.14)), we have that $U(x,t) := \Gamma_t * u(x)$, where

$$\Gamma_t(x) := G(x,t) = (4\pi t)^{-n/2} e^{-|x|^2/(4t)}.$$

We recall that the Fourier transform of a Gaussian is a Gaussian itself, namely

$$\mathscr{F}(e^{-\pi|x|^2}) = e^{-\pi|\xi|^2},$$

therefore, for any fixed $t > 0$, using the substitution $y := x/\sqrt{4\pi t}$,

$$\mathscr{F}\,\Gamma_t(\xi) = \frac{1}{(4\pi t)^{n/2}} \int_{\mathbb{R}^n} e^{-|x|^2/(4t)} e^{-2\pi i x\cdot\xi}\, dx$$

$$= \int_{\mathbb{R}^n} e^{-\pi|y|^2} e^{-2\pi i y\cdot(\sqrt{4\pi t}\,\xi)}\, dy$$

$$= e^{-4\pi^2 t|\xi|^2}.$$

As a consequence

$$\mathscr{F}\left(U(x,t)-u(x)\right) = \mathscr{F}\left(\Gamma_t * u(x) - u(x)\right)$$

$$= \mathscr{F}(\Gamma_t * u)(\xi) - \hat{u}(\xi) = \left(\mathscr{F}\,\Gamma_t(\xi)-1\right)\hat{u}(\xi)$$

$$= (e^{-4\pi^2 t|\xi|^2} - 1)\hat{u}(\xi).$$

We multiply by t^{-s-1} and integrate over $t > 0$, and we obtain

$$\mathscr{F} \int_0^{+\infty} t^{-s-1}\left(U(x,t)-u(x)\right) dt = \int_0^{+\infty} t^{-s-1}(e^{-4\pi^2 t|\xi|^2} - 1)\, dt\, \hat{u}(\xi)$$

$$= \Gamma(-s)\, (4\pi^2|\xi|^2)^s\, \hat{u}(\xi),$$

thanks to (2.11) (used here with $\lambda := 4\pi^2|\xi|^2$). By taking the inverse Fourier transform, we have

$$\int_0^{+\infty} t^{-s-1}\left(U(x,t)-u(x)\right) dt = \Gamma(-s)\, (2\pi)^{2s}\, \mathscr{F}^{-1}\left(|\xi|^{2s}\, \hat{u}(\xi)\right).$$

We insert this information into (2.19) and we get

$$\frac{2^{2s-1}\, s\, \Gamma\left(\frac{n}{2}+s\right)}{\pi^{n/2}\Gamma(1-s)} \int_{\mathbb{R}^n} \frac{2u(x)-u(x+y)-u(x-y)}{|y|^{n+2s}}\, dy = (2\pi)^{2s}\mathscr{F}^{-1}\big(|\xi|^{2s}\,\hat{u}(\xi)\big).$$

Hence, recalling (2.18),

$$\frac{2^{2s-1}\, s\, \Gamma\left(\frac{n}{2}+s\right)}{\pi^{n/2}\Gamma(1-s)} \int_{\mathbb{R}^n} \frac{2u(x)-u(x+y)-u(x-y)}{|y|^{n+2s}}\, dy$$

$$= \frac{1}{2\displaystyle\int_{\mathbb{R}^n}\frac{1-\cos\omega_1}{|\omega|^{n+2s}}\, d\omega} \int_{\mathbb{R}^n} \frac{2u(x)-u(x+y)-u(x-y)}{|y|^{n+2s}}\, dy,$$

which gives the desired result.

For the sake of completeness, a different proof of Lemma 2.3 will be given in Sect. A.2. There, to prove Lemma 2.3, we will use the theory of special functions rather than the fractional Laplacian. For other approaches to the proof of Lemma 2.3 see also the recent PhD dissertations [75] (and related [76]) and [92].

2.2 Fractional Sobolev Inequality and Generalized Coarea Formula

Fractional Sobolev spaces enjoy quite a number of important functional inequalities. It is almost impossible to list here all the results and the possible applications, therefore we will only present two important inequalities which have a simple and nice proof, namely the fractional Sobolev Inequality and the Generalized Coarea Formula.

The fractional Sobolev Inequality can be written as follows:

Theorem 2.2.1 *For any $s \in (0,1)$, $p \in \left(1, \frac{n}{s}\right)$ and $u \in C_0^\infty(\mathbb{R}^n)$,*

$$\|u\|_{L^{\frac{np}{n-sp}}(\mathbb{R}^n)} \leq C \left(\int_{\mathbb{R}^n}\int_{\mathbb{R}^n} \frac{|u(x)-u(y)|^p}{|x-y|^{n+sp}}\, dx\, dy \right)^{\frac{1}{p}}, \qquad (2.20)$$

for some $C > 0$, depending only on n and p.

Proof We follow here the very nice proof given in [117] (where, in turn, the proof is attributed to Haïm Brezis). We fix $r > 0$, $a > 0$, $P > 0$ and $x \in \mathbb{R}^n$. Then, for any $y \in \mathbb{R}^n$,

$$|u(x)| \leq |u(x)-u(y)| + |u(y)|,$$

and so, integrating over $B_r(x)$, we obtain

$$|B_r| \, |u(x)| \leq \int_{B_r(x)} |u(x) - u(y)| \, dy + \int_{B_r(x)} |u(y)| \, dy$$

$$= \int_{B_r(x)} \frac{|u(x) - u(y)|}{|x - y|^a} \cdot |x - y|^a \, dy + \int_{B_r(x)} |u(y)| \, dy$$

$$\leq r^a \int_{B_r(x)} \frac{|u(x) - u(y)|}{|x - y|^a} \, dy + \int_{B_r(x)} |u(y)| \, dy.$$

Now we choose $a := \frac{n+sp}{p}$ and we make use of the Hölder Inequality (with exponents p and $\frac{p}{p-1}$ and with exponents $\frac{np}{n-sp}$ and $\frac{np}{n(p-1)+sp}$), to obtain

$$|B_r| \, |u(x)| \leq r^{\frac{n+sp}{p}} \int_{B_r(x)} \frac{|u(x) - u(y)|}{|x - y|^{\frac{n+sp}{p}}} \, dy + \int_{B_r(x)} |u(y)| \, dy$$

$$\leq r^{\frac{n+sp}{p}} \left(\int_{B_r(x)} \frac{|u(x) - u(y)|^p}{|x - y|^{n+sp}} \, dy \right)^{\frac{1}{p}} \left(\int_{B_r(x)} dy \right)^{\frac{p-1}{p}}$$

$$+ \left(\int_{B_r(x)} |u(y)|^{\frac{np}{n-sp}} \, dy \right)^{\frac{n-sp}{np}} \left(\int_{B_r(x)} dy \right)^{\frac{n(p-1)+sp}{np}}$$

$$\leq C r^{n+s} \left(\int_{B_r(x)} \frac{|u(x) - u(y)|^p}{|x - y|^{n+sp}} \, dy \right)^{\frac{1}{p}}$$

$$+ C r^{\frac{n(p-1)+sp}{p}} \left(\int_{B_r(x)} |u(y)|^{\frac{np}{n-sp}} \, dy \right)^{\frac{n-sp}{np}},$$

for some $C > 0$. So, we divide by r^n and we possibly rename C. In this way, we obtain

$$|u(x)| \leq C r^s \left[\left(\int_{B_r(x)} \frac{|u(x) - u(y)|^p}{|x - y|^{n+sp}} \, dy \right)^{\frac{1}{p}} + r^{-\frac{n}{p}} \left(\int_{B_r(x)} |u(y)|^{\frac{np}{n-sp}} \, dy \right)^{\frac{n-sp}{np}} \right].$$

That is, using the short notation

$$\alpha := \int_{\mathbb{R}^n} \frac{|u(x) - u(y)|^p}{|x - y|^{n+sp}} \, dy$$

$$\text{and} \quad \beta := \int_{\mathbb{R}^n} |u(y)|^{\frac{np}{n-sp}} \, dy$$

we have that

$$|u(x)| \leq C r^s \left(\alpha^{\frac{1}{p}} + r^{-\frac{n}{p}} \beta^{\frac{n-sp}{np}} \right),$$

hence, raising both terms at the appropriate power $\frac{np}{n-sp}$ and renaming C

$$|u(x)|^{\frac{np}{n-sp}} \leq Cr^{\frac{nsp}{n-sp}} \left(\alpha^{\frac{1}{p}} + r^{-\frac{n}{p}} \beta^{\frac{n-sp}{np}} \right)^{\frac{np}{n-sp}}. \tag{2.21}$$

We take now

$$r := \frac{\beta^{\frac{n-sp}{n^2}}}{\alpha^{\frac{1}{n}}}.$$

With this setting, we have that $r^{-\frac{n}{p}} \beta^{\frac{n-sp}{np}}$ is equal to $\alpha^{\frac{1}{p}}$. Accordingly, possibly renaming C, we infer from (2.21) that

$$|u(x)|^{\frac{np}{n-sp}} \leq C\alpha \, \beta^{\frac{sp}{n}}$$

$$= C \int_{\mathbb{R}^n} \frac{|u(x)-u(y)|^p}{|x-y|^{n+sp}} \, dy \left(\int_{\mathbb{R}^n} |u(y)|^{\frac{np}{n-sp}} \, dy \right)^{\frac{sp}{n}},$$

for some $C > 0$, and so, integrating over $x \in \mathbb{R}^n$,

$$\int_{\mathbb{R}^n} |u(x)|^{\frac{np}{n-sp}} \, dx \leq C \left(\int_{\mathbb{R}^n} \int_{\mathbb{R}^n} \frac{|u(x)-u(y)|^p}{|x-y|^{n+sp}} \, dx \, dy \right) \left(\int_{\mathbb{R}^n} |u(y)|^{\frac{np}{n-sp}} \, dy \right)^{\frac{sp}{n}}.$$

This, after a simplification, gives (2.20).

What follows is the Generalized Co-area Formula of [139] (the link with the classical Co-area Formula will be indeed more evident in terms of the fractional perimeter functional discussed in Chap. 5).

Theorem 2.2.2 *For any $s \in (0,1)$ and any measurable function $u : \Omega \to [0,1]$,*

$$\frac{1}{2} \int_\Omega \int_\Omega \frac{|u(x)-u(y)|}{|x-y|^{n+s}} \, dx \, dy = \int_0^1 \left(\int_{\{x \in \Omega, \, u(x)>t\}} \int_{\{y \in \Omega, \, u(y) \leq t\}} \frac{dx \, dy}{|x-y|^{n+s}} \right) dt.$$

Proof We claim that for any $x, y \in \Omega$

$$|u(x) - u(y)| = \int_0^1 \Big(\chi_{\{u>t\}}(x) \, \chi_{\{u \leq t\}}(y) + \chi_{\{u \leq t\}}(x) \, \chi_{\{u>t\}}(y) \Big) \, dt. \tag{2.22}$$

To prove this, we fix x and y in Ω, and by possibly exchanging them, we can suppose that $u(x) \geq u(y)$. Then, we define

$$\varphi(t) := \chi_{\{u>t\}}(x) \, \chi_{\{u \leq t\}}(y) + \chi_{\{u \leq t\}}(x) \, \chi_{\{u>t\}}(y).$$

By construction

$$\varphi(t) = \begin{cases} 0 & \text{if} \quad t < u(y) \text{ and } t \geq u(x), \\ 1 & \text{if} \quad u(y) \leq t < u(x), \end{cases}$$

therefore

$$\int_0^1 \varphi(t)\, dt = \int_{u(y)}^{u(x)} dt = u(x) - u(y),$$

which proves (2.22).

So, multiplying by the singular kernel and integrating (2.22) over $\Omega \times \Omega$, we obtain that

$$\int_\Omega \int_\Omega \frac{|u(x) - u(y)|}{|x - y|^{n+s}}\, dx\, dy$$

$$= \int_0^1 \left(\int_\Omega \int_\Omega \frac{\chi_{\{u>t\}}(x)\, \chi_{\{u\leq t\}}(y) + \chi_{\{u\leq t\}}(x)\, \chi_{\{u>t\}}(y)}{|x - y|^{n+s}}\, dx\, dy \right) dt$$

$$= \int_0^1 \left(\int_{\{u>t\}} \int_{\{u\leq t\}} \frac{dx\, dy}{|x - y|^{n+s}} + \int_{\{u\leq t\}} \int_{\{u>t\}} \frac{dx\, dy}{|x - y|^{n+s}} \right) dt$$

$$= 2 \int_0^1 \left(\int_{\{u>t\}} \int_{\{u\leq t\}} \frac{dx\, dy}{|x - y|^{n+s}} \right) dt,$$

as desired.

2.3 Maximum Principle and Harnack Inequality

The Harnack Inequality and the Maximum Principle for harmonic functions are classical topics in elliptic regularity theory. Namely, in the classical case, if a non-negative function is harmonic in B_1 and $r \in (0, 1)$, then its minimum and maximum in B_r must always be comparable (in particular, the function cannot touch the level zero in B_r).

It is worth pointing out that the fractional counterpart of these facts is, in general, false, as this next simple result shows (see [94]):

Theorem 2.3.1 *There exists a bounded function u which is s-harmonic in B_1, non-negative in B_1, but such that* $\inf_{B_1} u = 0$.

Proof (Sketch of the proof) The main idea is that we are able to take the datum of u outside B_1 in a suitable way as to "bend down" the function inside B_1 until it reaches the level zero. Namely, let $M \geq 0$ and we take u_M to be the function satisfying

$$\begin{cases} (-\Delta)^s u_M = 0 & \text{in } B_1, \\ u_M = 1 - M & \text{in } B_3 \setminus B_2, \\ u_M = 1 & \text{in } \mathbb{R}^n \setminus \big((B_3 \setminus B_2) \cup B_1\big). \end{cases} \tag{2.23}$$

When $M = 0$, the function u_M is identically 1. When $M > 0$, we expect u_M to bend down, since the fact that the fractional Laplacian vanishes in B_1 forces the second order quotient to vanish in average (recall (1.1), or the equivalent formulation in (2.5)). Indeed, we claim that there exists $M_\star > 0$ such that $u_{M_\star} \geq 0$ in B_1 with $\inf_{B_1} u_{M_\star} = 0$. Then, the result of Theorem 2.3.1 would be reached by taking $u := u_{M_\star}$.

To check the existence of such $M_\star$, we show that $\inf_{B_1} u_M \to -\infty$ as $M \to +\infty$. Indeed, we argue by contradiction and suppose this cannot happen. Then, for any $M \geq 0$, we would have that

$$\inf_{B_1} u_M \geq -a, \tag{2.24}$$

for some fixed $a \in \mathbb{R}$. We set

$$v_M := \frac{u_M + M - 1}{M}.$$

Then, by (2.23),

$$\begin{cases} (-\Delta)^s v_M = 0 & \text{in } B_1, \\ v_M = 0 & \text{in } B_3 \setminus B_2, \\ v_M = 1 & \text{in } \mathbb{R}^n \setminus \big((B_3 \setminus B_2) \cup B_1\big). \end{cases}$$

Also, by (2.24), for any $x \in B_1$,

$$v_M(x) \geq \frac{-a + M - 1}{M}.$$

By taking limits, one obtains that v_M approaches a function v_∞ that satisfies

$$\begin{cases} (-\Delta)^s v_\infty = 0 & \text{in } B_1, \\ v_\infty = 0 & \text{in } B_3 \setminus B_2, \\ v_\infty = 1 & \text{in } \mathbb{R}^n \setminus \big((B_3 \setminus B_2) \cup B_1\big) \end{cases}$$

and, for any $x \in B_1$,

$$v_\infty(x) \geq 1.$$

In particular the maximum of v_∞ is attained at some point $x_\star \in B_1$, with $v_\infty(x_\star) \geq 1$. Accordingly,

$$0 = P.V. \int_{\mathbb{R}^n} \frac{v_\infty(x_\star) - v_\infty(y)}{|x_\star - y|^{n+2s}} \, dy \geq P.V. \int_{B_3 \setminus B_2} \frac{v_\infty(x_\star) - v_\infty(y)}{|x_\star - y|^{n+2s}} \, dy$$

$$\geq P.V. \int_{B_3 \setminus B_2} \frac{1 - 0}{|x_\star - y|^{n+2s}} \, dy > 0,$$

which is a contradiction.

The example provided by Theorem 2.3.1 is not the end of the story concerning the Harnack Inequality in the fractional setting. On the one hand, Theorem 2.3.1 is just a particular case of the very dramatic effect that the datum at infinity may have on the fractional Laplacian (a striking example of this phenomenon will be given in Sect. 2.5). On the other hand, the Harnack Inequality and the Maximum Principle hold true if, for instance, the sign of the function u is controlled in the whole of $\mathbb{R}^n$.

We refer to [11, 77, 94, 130] and to the references therein for a detailed introduction to the fractional Harnack inequality, and to [55] for general estimates of this type.

Just to point out the validity of a global Maximum Principle, we state in detail the following simple result:

Theorem 2.3.2 *If* $(-\Delta)^s u \geq 0$ *in* B_1 *and* $u \geq 0$ *in* $\mathbb{R}^n \setminus B_1$*, then* $u \geq 0$ *in* B_1.

Proof Suppose by contradiction that the minimal point $x_\star \in B_1$ satisfies $u(x_\star) < 0$. Then $u(x_\star)$ is a minimum in $\mathbb{R}^n$ (since u is positive outside B_1), if $y \in B_2$ we have that $2u(x_\star) - u(x_\star + y) - u(x_\star - y) \leq 0$. On the other hand, in $\mathbb{R}^n \setminus B_2$ we have that $x_\star \pm y \in \mathbb{R}^n \setminus B_1$, hence $u(x_\star \pm y) \geq 0$. We thus have

$$0 \leq \int_{\mathbb{R}^n} \frac{2u(x_\star) - u(x_\star + y) - u(x_\star - y)}{|y|^{n+2s}} \, dy$$

$$\leq \int_{\mathbb{R}^n \setminus B_2} \frac{2u(x_\star) - u(x_\star + y) - u(x_\star - y)}{|y|^{n+2s}} \, dy$$

$$\leq \int_{\mathbb{R}^n \setminus B_2} \frac{2u(x_\star)}{|y|^{n+2s}} \, dy < 0.$$

This leads to a contradiction.

Similarly to Theorem 2.3.2, one can prove a Strong Maximum Principle, such as:

Theorem 2.3.3 *If* $(-\Delta)^s u \geq 0$ *in* B_1 *and* $u \geq 0$ *in* $\mathbb{R}^n \setminus B_1$*, then* $u > 0$ *in* B_1*, unless u vanishes identically.*

Proof We observe that we already know that $u \geq 0$ in the whole of $\mathbb{R}^n$, thanks to Theorem 2.3.2. Hence, if u is not strictly positive, there exists $x_0 \in B_1$ such that $u(x_0) = 0$. This gives that

$$0 \leq \int_{\mathbb{R}^n} \frac{2u(x_0) - u(x_0 + y) - u(x_0 - y)}{|y|^{n+2s}} \, dy = -\int_{\mathbb{R}^n} \frac{u(x_0 + y) + u(x_0 - y)}{|y|^{n+2s}} \, dy.$$

Now both $u(x_0 + y)$ and $u(x_0 - y)$ are non-negative, hence the latter integral is less than or equal to zero, and so it must vanish identically, proving that u also vanishes identically.

A simple version of a Harnack-type inequality in the fractional setting can be also obtained as follows:

Proposition 2.3.4 *Assume that $(-\Delta)^s u \geq 0$ in B_2, with $u \geq 0$ in the whole of $\mathbb{R}^n$. Then*

$$u(0) \geq c \int_{B_1} u(x) \, dx,$$

for a suitable $c > 0$.

Proof Let $\Gamma \in C_0^\infty(B_{1/2})$, with $\Gamma(x) \in [0, 1]$ for any $x \in \mathbb{R}^n$, and $\Gamma(0) = 1$. We fix $\epsilon > 0$, to be taken arbitrarily small at the end of this proof and set

$$\eta := u(0) + \epsilon > 0. \tag{2.25}$$

We define $\Gamma_a(x) := 2\eta \, \Gamma(x) - a$. Notice that if $a > 2\eta$, then $\Gamma_a(x) \leq 2\eta - a < 0 \leq u(x)$ in the whole of $\mathbb{R}^n$, hence the set $\{\Gamma_a < u \text{ in } \mathbb{R}^n\}$ is not empty, and we can define

$$a_* := \inf_{a \in \mathbb{R}} \{\Gamma_a < u \text{ in } \mathbb{R}^n\}.$$

By construction

$$a_* \leq 2\eta. \tag{2.26}$$

If $a < \eta$ then $\Gamma_a(0) = 2\eta - a > \eta > u(0)$, therefore

$$a_* \geq \eta. \tag{2.27}$$

Notice that

$$\Gamma_{a_*} \leq u \text{ in the whole of } \mathbb{R}^n. \tag{2.28}$$

We claim that

$$\text{there exists } x_0 \in \overline{B_{1/2}} \text{ such that } \Gamma_{a_*}(x_0) = u(x_0). \tag{2.29}$$

To prove this, we suppose by contradiction that $u > \Gamma_{a_*}$ in $\overline{B_{1/2}}$, i.e.

$$\mu := \min_{\overline{B_{1/2}}} (u - \Gamma_{a_*}) > 0.$$

Also, if $x \in \mathbb{R}^n \setminus \overline{B_{1/2}}$, we have that

$$u(x) - \Gamma_{a_*}(x) = u(x) - 2\eta\,\Gamma(x) + a_* = u(x) + a_* \geq a_* \geq \eta,$$

thanks to (2.27). As a consequence, for any $x \in \mathbb{R}^n$,

$$u(x) - \Gamma_{a_*}(x) \geq \min\{\mu, \eta\} =: \mu_* > 0.$$

So, if we define $a_\sharp := a_* - (\mu_*/2)$, we have that $a_\sharp < a_*$ and

$$u(x) - \Gamma_{a_\sharp}(x) = u(x) - \Gamma_{a_*}(x) - \frac{\mu_*}{2} \geq \frac{\mu_*}{2} > 0.$$

This is in contradiction with the definition of a_* and so it proves (2.29).

From (2.29) we have that $x_0 \in \overline{B_{1/2}}$, hence $(-\Delta)^s u(x_0) \geq 0$. Also $|(-\Delta)^s \Gamma_{a_*}(x)|$ $= 2\eta\,|(-\Delta)^s \Gamma(x)| \leq C\eta$, for any $x \in \mathbb{R}^n$, therefore, recalling (2.28) and (2.29),

$$C\eta \geq (-\Delta)^s \Gamma_{a_*}(x_0) - (-\Delta)^s u(x_0)$$

$$= C(n,s)\,\text{P.V.}\int_{\mathbb{R}^n} \frac{\left[\Gamma_{a_*}(x_0) - \Gamma_{a_*}(x_0 + y)\right] - \left[u(x_0) - u(x_0 + y)\right]}{|y|^{n+2s}}\,dy$$

$$= C(n,s)\,\text{P.V.}\int_{\mathbb{R}^n} \frac{u(x_0 + y) - \Gamma_{a_*}(x_0 + y)}{|y|^{n+2s}}\,dy$$

$$\geq C(n,s)\,\text{P.V.}\int_{B_1(-x_0)} \frac{u(x_0 + y) - \Gamma_{a_*}(x_0 + y)}{|y|^{n+2s}}\,dy.$$

Notice now that if $y \in B_1(-x_0)$, then $|y| \leq |x_0| + 1 < 2$, thus we obtain

$$C\eta \geq \frac{C(n,s)}{2^{n+2s}}\int_{B_1(-x_0)} \left[u(x_0 + y) - \Gamma_{a_*}(x_0 + y)\right]dy.$$

Notice now that $\Gamma_{a_*}(x) = 2\eta\Gamma(x) - a_* \leq \eta$, due to (2.27), therefore we conclude that

$$C\eta \geq \frac{C(n,s)}{2^{n+2s}}\left(\int_{B_1(-x_0)} u(x_0 + y)\,dy - \eta|B_1|\right).$$

That is, using the change of variable $x := x_0 + y$, recalling (2.25) and renaming the constants, we have

$$C\big(u(0) + \epsilon\big) = C\eta \geq \int_{B_1} u(x)\,dx,$$

hence the desired result follows by sending $\epsilon \to 0$.

2.4 An s-Harmonic Function

We provide here an explicit example of a function that is s-harmonic on the positive line $\mathbb{R}_+ := (0, +\infty)$. Namely, we prove the following result:

Theorem 2.4.1 *For any $x \in \mathbb{R}$, let $w_s(x) := x_+^s = \max\{x, 0\}^s$. Then*

$$(-\Delta)^s w_s(x) = \begin{cases} -c_s |x|^{-s} & \text{if } x < 0, \\ 0 & \text{if } x > 0, \end{cases}$$

for a suitable constant $c_s > 0$.

At a first glance, it may be quite surprising that the function x_+^s is s-harmonic in $(0, +\infty)$, since such function is not smooth (but only continuous) uniformly up to the boundary, so let us try to give some heuristic explanations for it (Fig. 2.1).

We try to understand why the function x_+^s is s-harmonic in, say, the interval $(0, 1)$ when $s \in (0, 1]$. From the discussion in Sect. 1.2, we know that the s-harmonic function in $(0, 1)$ that agrees with x_+^s outside $(0, 1)$ coincides with the expected value of a payoff, subject to a random walk (the random walk is classical when $s = 1$ and it presents jumps when $s \in (0, 1)$). If $s = 1$ and we start from the middle of the interval, we have the same probability of being moved randomly to the left and to the right. This means that we have the same probability of exiting the interval $(0, 1)$ to the right (and so ending the process at $x = 1$, which gives 1 as payoff) or to the left (and so ending the process at $x = 0$, which gives 0 as payoff). Therefore the expected value starting at $x = 1/2$ is exactly the average between 0 and 1, which is $1/2$. Similarly, if we start the process at the point $x = 1/4$, we have the same probability of reaching the point 0 on the left and the point $1/2$ to the right. Since

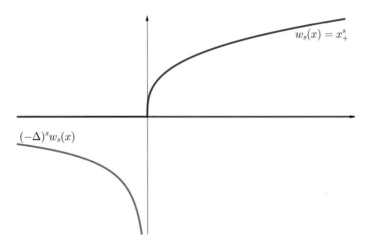

Fig. 2.1 An s-harmonic function

we know that the payoff at $x = 0$ is 0 and the expected value of the payoff at $x = 1/2$ is $1/2$, we deduce in this case that the expected value for the process starting at $1/4$ is the average between 0 and $1/2$, that is exactly $1/4$. We can repeat this argument over and over, and obtain the (rather obvious) fact that the linear function is indeed harmonic in the classical sense.

The argument above, which seems either trivial or unnecessarily complicated in the classical case, can be adapted when $s \in (0, 1)$ and it can give a qualitative picture of the corresponding *s*-harmonic function. Let us start again the random walk, this time with jumps, at $x = 1/2$: in presence of jumps, we have the same probability of reaching the left interval $(-\infty, 0]$ and the right interval $[1, +\infty)$. Now, the payoff at $(-\infty, 0]$ is 0, while the payoff at $[1, +\infty)$ is *bigger* than 1. This implies that the expected value at $x = 1/2$ is the average between 0 and something bigger than 1, which produces a value larger than $1/2$. When repeating this argument over and over, we obtain a concavity property enjoyed by the *s*-harmonic functions in this case (the exact values prescribed in $[1, +\infty)$ are not essential here, it would be enough that these values were monotone increasing and larger than 1) (Fig. 2.2).

In a sense, therefore, this concavity properties and loss of Lipschitz regularity near the minimal boundary values is typical of nonlocal diffusion and it is due to the possibility of "reaching far away points", which may increase the expected payoff.

Now we present a complete, elementary proof of Theorem 2.4.1. The proof originated from some pleasant discussions with Fernando Soria and it is based on some rather surprising integral cancellations. The reader who wishes to skip this proof can go directly to Sect. 2.3 on page 19. Moreover, a shorter, but technically more advanced proof, is presented in Sect. A.1.

Here, we start with some preliminary computations.

Lemma 2.4 *For any* $s \in (0, 1)$

$$\int_0^1 \frac{(1+t)^s + (1-t)^s - 2}{t^{1+2s}}\, dt + \int_1^{+\infty} \frac{(1+t)^s}{t^{1+2s}}\, dt = \frac{1}{s}.$$

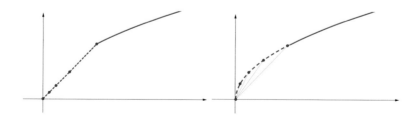

Fig. 2.2 A payoff model: case $s = 1$ and $s \in (0, 1)$

Proof Fixed $\varepsilon > 0$, we integrate by parts:

$$\int_\varepsilon^1 \frac{(1+t)^s + (1-t)^s - 2}{t^{1+2s}}\, dt$$

$$= -\frac{1}{2s}\int_\varepsilon^1 \left[(1+t)^s + (1-t)^s - 2\right]\frac{d}{dt} t^{-2s}\, dt$$

$$= \frac{1}{2s}\left[\frac{(1+\varepsilon)^s + (1-\varepsilon)^s - 2}{\varepsilon^{2s}} - 2^s + 2\right] + \frac{1}{2}\int_\varepsilon^1 \frac{(1+t)^{s-1} - (1-t)^{s-1}}{t^{2s}}\, dt$$

$$= \frac{1}{2s}\left[o(1) - 2^s + 2\right] + \frac{1}{2}\left(\int_\varepsilon^1 (1+t)^{s-1} t^{-2s}\, dt - \int_\varepsilon^1 (1-t)^{s-1} t^{-2s}\, dt\right),$$

$$(2.30)$$

with $o(1)$ infinitesimal as $\varepsilon \searrow 0$. Moreover, by changing variable $\tilde{t} := t/(1-t)$, that is $t := \tilde{t}/(1+\tilde{t})$, we have that

$$\int_\varepsilon^1 (1-t)^{s-1} t^{-2s}\, dt = \int_{\varepsilon/(1-\varepsilon)}^{+\infty} (1+\tilde{t})^{s-1}\tilde{t}^{-2s}\, d\tilde{t}.$$

Inserting this into (2.30) (and writing t instead of $\tilde{t}$ as variable of integration), we obtain

$$\int_\varepsilon^1 \frac{(1+t)^s + (1-t)^s - 2}{t^{1+2s}}\, dt$$

$$= \frac{1}{2s}\left[o(1) - 2^s + 2\right] + \frac{1}{2}\left[\int_\varepsilon^1 (1+t)^{s-1} t^{-2s}\, dt - \int_{\varepsilon/(1-\varepsilon)}^{+\infty} (1+t)^{s-1} t^{-2s}\, dt\right]$$

$$= \frac{1}{2s}\left[o(1) - 2^s + 2\right] + \frac{1}{2}\left[\int_\varepsilon^{\varepsilon/(1-\varepsilon)} (1+t)^{s-1} t^{-2s}\, dt - \int_1^{+\infty} (1+t)^{s-1} t^{-2s}\, dt\right].$$

$$(2.31)$$

Now we remark that

$$\int_\varepsilon^{\varepsilon/(1-\varepsilon)} (1+t)^{s-1} t^{-2s}\, dt \le \int_\varepsilon^{\varepsilon/(1-\varepsilon)} (1+\varepsilon)^{s-1}\varepsilon^{-2s}\, dt = \varepsilon^{2-2s}(1-\varepsilon)^{-1}(1+\varepsilon)^{s-1},$$

therefore

$$\lim_{\varepsilon \searrow 0} \int_\varepsilon^{\varepsilon/(1-\varepsilon)} (1+t)^{s-1} t^{-2s}\, dt = 0.$$

So, by passing to the limit in (2.31), we get

$$\int_0^1 \frac{(1+t)^s + (1-t)^s - 2}{t^{1+2s}}\, dt = \frac{-2^s + 2}{2s} - \frac{1}{2}\int_1^{+\infty} (1+t)^{s-1} t^{-2s}\, dt. \qquad (2.32)$$

Now, integrating by parts we see that

$$
\frac{1}{2} \int_1^{+\infty} (1+t)^{s-1} t^{-2s}\, dt = \frac{1}{2s} \int_1^{+\infty} t^{-2s} \frac{d}{dt} (1+t)^s\, dt
$$

$$
= -\frac{2^s}{2s} + \int_1^{+\infty} t^{-1-2s}(1+t)^s\, dt.
$$

By plugging this into (2.32) we obtain that

$$
\int_0^1 \frac{(1+t)^s + (1-t)^s - 2}{t^{1+2s}}\, dt = \frac{-2^s + 2}{2s} + \frac{2^s}{2s} - \int_1^{+\infty} t^{-1-2s}(1+t)^s\, dt,
$$

which gives the desired result.

From Lemma 2.4 we deduce the following (somehow unexpected) cancellation property:

Corollary 2.4.1 *Let w_s be as in the statement of Theorem 2.4.1. Then*

$$
(-\Delta)^s w_s(1) = 0.
$$

Proof The function $t \mapsto (1+t)^s + (1-t)^s - 2$ is even, therefore

$$
\int_{-1}^1 \frac{(1+t)^s + (1-t)^s - 2}{|t|^{1+2s}}\, dt = 2 \int_0^1 \frac{(1+t)^s + (1-t)^s - 2}{t^{1+2s}}\, dt.
$$

Moreover, by changing variable $\tilde{t} := -t$, we have that

$$
\int_{-\infty}^{-1} \frac{(1-t)^s - 2}{|t|^{1+2s}}\, dt = \int_1^{+\infty} \frac{(1+\tilde{t})^s - 2}{\tilde{t}^{1+2s}}\, d\tilde{t}.
$$

Therefore

$$
\int_{-\infty}^{+\infty} \frac{w_s(1+t) + w_s(1-t) - 2w_s(1)}{|t|^{1+2s}}\, dt
$$

$$
= \int_{-\infty}^{-1} \frac{(1-t)^s - 2}{|t|^{1+2s}}\, dt + \int_{-1}^1 \frac{(1+t)^s + (1-t)^s - 2}{|t|^{1+2s}}\, dt + \int_1^{+\infty} \frac{(1+t)^s - 2}{|t|^{1+2s}}\, dt
$$

$$
= 2 \int_0^1 \frac{(1+t)^s + (1-t)^s - 2}{t^{1+2s}}\, dt + 2 \int_1^{+\infty} \frac{(1+t)^s - 2}{t^{1+2s}}\, dt
$$

$$
= 2 \left[\int_0^1 \frac{(1+t)^s + (1-t)^s - 2}{t^{1+2s}}\, dt + \int_1^{+\infty} \frac{(1+t)^s}{t^{1+2s}}\, dt - 2 \int_1^{+\infty} \frac{dt}{t^{1+2s}} \right]
$$

$$
= 2 \left[\frac{1}{s} - 2 \int_1^{+\infty} \frac{dt}{t^{1+2s}} \right],
$$

where Lemma 2.4 was used in the last line. Since

$$\int_{1}^{+\infty} \frac{dt}{t^{1+2s}} = \frac{1}{2s},$$

we obtain that

$$\int_{-\infty}^{+\infty} \frac{w_s(1+t) + w_s(1-t) - 2w_s(1)}{|t|^{1+2s}}\, dt = 0,$$

that proves the desired claim.

The counterpart of Corollary 2.4.1 is given by the following simple observation:

Lemma 2.5 *Let w_s be as in the statement of Theorem 2.4.1. Then*

$$-(-\Delta)^s w_s(-1) > 0.$$

Proof We have that

$$w_s(-1+t) + w_s(-1-t) - 2w_s(-1) = (-1+t)_+^s + (-1-t)_+^s \geq 0$$

and not identically zero, which implies the desired result.

We have now all the elements to proceed to the proof of Theorem 2.4.1.

Proof (Proof of Theorem 2.4.1)
   We let $\sigma \in \{+1, -1\}$ denote the sign of a fixed $x \in \mathbb{R} \setminus \{0\}$. We claim that

$$\int_{-\infty}^{+\infty} \frac{w_s(\sigma(1+t)) + w_s(\sigma(1-t)) - 2w_s(\sigma)}{|t|^{1+2s}}\, dt$$
$$= \int_{-\infty}^{+\infty} \frac{w_s(\sigma+t) + w_s(\sigma-t) - 2w_s(\sigma)}{|t|^{1+2s}}\, dt. \tag{2.33}$$

Indeed, the formula above is obvious when $x > 0$ (i.e. $\sigma = 1$), so we suppose $x < 0$ (i.e. $\sigma = -1$) and we change variable $\tau := -t$, to see that, in this case,

$$\int_{-\infty}^{+\infty} \frac{w_s(\sigma(1+t)) + w_s(\sigma(1-t)) - 2w_s(\sigma)}{|t|^{1+2s}}\, dt$$
$$= \int_{-\infty}^{+\infty} \frac{w_s(-1-t) + w_s(-1+t) - 2w_s(\sigma)}{|t|^{1+2s}}\, dt$$
$$= \int_{-\infty}^{+\infty} \frac{w_s(-1+\tau) + w_s(-1-\tau) - 2w_s(\sigma)}{|\tau|^{1+2s}}\, d\tau$$
$$= \int_{-\infty}^{+\infty} \frac{w_s(\sigma+\tau) + w_s(\sigma-\tau) - 2w_s(\sigma)}{|\tau|^{1+2s}}\, d\tau,$$

thus checking (2.33).

Now we observe that, for any $r \in \mathbb{R}$,

$$w_s(|x|r) = (|x|r)_+^s = |x|^s r_+^s = |x|^s w_s(r).$$

That is

$$w_s(xr) = w_s(\sigma|x|r) = |x|^s w_s(\sigma r).$$

So we change variable $y = tx$ and we obtain that

$$\int_{-\infty}^{+\infty} \frac{w_s(x+y) + w_s(x-y) - 2w_s(x)}{|y|^{1+2s}}\, dy$$

$$= \int_{-\infty}^{+\infty} \frac{w_s(x(1+t)) + w_s(x(1-t)) - 2w_s(x)}{|x|^{2s}|t|^{1+2s}}\, dt$$

$$= |x|^{-s} \int_{-\infty}^{+\infty} \frac{w_s(\sigma(1+t)) + w_s(\sigma(1-t)) - 2w_s(\sigma)}{|t|^{1+2s}}\, dt$$

$$= |x|^{-s} \int_{-\infty}^{+\infty} \frac{w_s(\sigma+t) + w_s(\sigma-t) - 2w_s(\sigma)}{|t|^{1+2s}}\, dt,$$

where (2.33) was used in the last line. This says that

$$(-\Delta)^s w_s(x) = \begin{cases} |x|^{-s}\,(-\Delta)^s w_s(-1) & \text{if } x < 0, \\ |x|^{-s}\,(-\Delta)^s w_s(1) & \text{if } x > 0, \end{cases}$$

hence the result in Theorem 2.4.1 follows from Corollary 2.4.1 and Lemma 2.5.

2.5 All Functions Are Locally s-Harmonic Up to a Small Error

Here we will show that s-harmonic functions can locally approximate any given function, without any geometric constraints. This fact is rather surprising and it is a purely nonlocal feature, in the sense that it has no classical counterpart. Indeed, in the classical setting, harmonic functions are quite rigid, for instance they cannot have a strict local maximum, and therefore cannot approximate a function with a strict local maximum. The nonlocal picture is, conversely, completely different, as the oscillation of a function "from far" can make the function locally harmonic, almost independently from its local behavior.

We want to give here some hints on the proof of this approximation result:

Theorem 2.5.1 *Let $k \in \mathbb{N}$ be fixed. Then for any $f \in C^k(\overline{B_1})$ and any $\varepsilon > 0$ there exists $R > 0$ and $u \in H^s(\mathbb{R}^n) \cap C^s(\mathbb{R}^n)$ such that*

$$\begin{cases} (-\Delta)^s u(x) = 0 & \text{in } B_1 \\ u = 0 & \text{in } \mathbb{R}^n \setminus B_R \end{cases} \tag{2.34}$$

and

$$\|f - u\|_{C^k(\overline{B_1})} \leq \varepsilon.$$

Proof (Sketch of the proof) For the sake of convenience, we divide the proof into three steps. Also, for simplicity, we give the sketch of the proof in the one-dimensional case. See [62] for the entire and more general proof.

Step 1. Reducing to monomials
Let $k \in \mathbb{N}$ be fixed. We use first of all the Stone-Weierstrass Theorem and we have that for any $\varepsilon > 0$ and any $f \in C^k([0, 1])$ there exists a polynomial P such that

$$\|f - P\|_{C^k(\overline{B_1})} \leq \varepsilon.$$

Hence it is enough to prove Theorem 2.5.1 for polynomials. Then, by linearity, it is enough to prove it for monomials. Indeed, if $P(x) = \sum_{m=0}^{N} c_m x^m$ and one finds an s-harmonic function u_m such that

$$\|u_m - x^m\|_{C^k(\overline{B_1})} \leq \frac{\varepsilon}{|c_m|},$$

then by taking $u := \sum_{m=1}^{N} c_m u_m$ we have that

$$\|u - P\|_{C^k(\overline{B_1})} \leq \sum_{m=1}^{N} |c_m| \|u_m - x^m\|_{C^k(\overline{B_1})} \leq \varepsilon.$$

Notice that the function u is still s-harmonic, since the fractional Laplacian is a linear operator.

Step 2. Spanning the derivatives
We prove the existence of an s-harmonic function in B_1, vanishing outside a compact set and with arbitrarily large number of derivatives prescribed. That is, we show that for any $m \in \mathbb{N}$ there exist $R > r > 0$, a point $x \in \mathbb{R}$ and a function u such that

$$(-\Delta)^s u = 0 \text{ in } (x - r, x + r),$$
$$u = 0 \text{ outside } (x - R, x + R), \tag{2.35}$$

and

$$D^j u(x) = 0 \text{ for any } j \in \{0, \ldots, m-1\},$$
$$D^m u(x) = 1.$$
(2.36)

To prove this, we argue by contradiction.

We consider $\mathscr{Z}$ to be the set of all pairs (u, x) of *s*-harmonic functions in a neighborhood of x, and points $x \in \mathbb{R}$ satisfying (2.35). To any pair, we associate the vector

$$\big(u(x), Du(x), \ldots, D^m u(x)\big) \in \mathbb{R}^{m+1}$$

and take V to be the vector space spanned by this construction, i.e.

$$V := \Big\{ \big(u(x), Du(x), \ldots, D^m u(x)\big), \text{ for } (u, x) \in \mathscr{Z} \Big\}.$$

Notice indeed that

$$V \text{ is a linear space.}$$
(2.37)

Indeed, let $V_1, V_2 \in V$ and $a_1, a_2 \in \mathbb{R}$. Then, for any $i \in \{1, 2\}$, we have that $V_i = \big(u_i(x_i), Du_i(x_i), \ldots, D^m u_i(x_i)\big)$, for some $(u_i, x_i) \in \mathscr{Z}$, i.e. u_i is *s*-harmonic in $(x_i - r_i, x_i + r_i)$ and vanishes outside $(x_i - R_i, x_i + R_i)$, for some $R_i \geq r_i > 0$. We set

$$u_3(x) := a_1 u_1(x + x_1) + a_2 u_2(x + x_2).$$

By construction, u_3 is *s*-harmonic in $(-r_3, r_3)$, and it vanishes outside $(-R_3, R_3)$, with $r_3 := \min\{r_1, r_2\}$ and $R_3 := \max\{R_1, R_2\}$, therefore $(u_3, 0) \in \mathscr{Z}$. Moreover

$$D^j u_3(x) = a_1 D^j u_1(x + x_1) + a_2 D^j u_2(x + x_2)$$

and thus

$$\begin{aligned}
&a_1 V_1 + a_2 V_2 \\
&= a_1 \big(u_1(x_1), Du_1(x_1), \ldots, D^m u_1(x_1)\big) + a_2 \big(u_2(x_2), Du_2(x_2), \ldots, D^m u_2(x_2)\big) \\
&= \big(u_3(0), Du_3(0), \ldots, D^m u_3(0)\big).
\end{aligned}$$

This establishes (2.37).

Now, to complete the proof of Step 2, it is enough to show that

$$V = \mathbb{R}^{m+1}.$$
(2.38)

Indeed, if (2.38) holds true, then in particular $(0, \ldots, 0, 1) \in V$, which is the desired claim in Step 2.

To prove (2.38), we argue by contradiction: if not, by (2.37), we have that V is a proper subspace of $\mathbb{R}^{m+1}$ and so it lies in a hyperplane.

Hence there exists a vector $c = (c_0, \ldots, c_m) \in \mathbb{R}^{m+1} \setminus \{0\}$ such that

$$V \subseteq \{\zeta \in \mathbb{R}^{m+1} \text{ s.t. } c \cdot \zeta = 0\}.$$

That is, taking a pair $(u, x) \in \mathscr{Z}$, the vector $c = (c_0, \ldots, c_m)$ is orthogonal to any vector $\big(u(x), Du(x), \ldots, D^m u(x)\big)$, namely

$$\sum_{j \le m} c_j D^j u(x) = 0.$$

To find a contradiction, we now choose an appropriate s-harmonic function u and we evaluate it at an appropriate point x. As a matter of fact, a good candidate for the s-harmonic function is x_+^s, as we know from Theorem 2.4.1: strictly speaking, this function is not allowed here, since it is not compactly supported, but let us say that one can construct a compactly supported s-harmonic function with the same behavior near the origin. With this slight caveat set aside, we compute for a (possibly small) x in $(0, 1)$:

$$D^j x^s = s(s-1) \ldots (s-j+1) x^{s-j}$$

and multiplying the sum with x^{m-s} (for $x \ne 0$) we have that

$$\sum_{j \le m} c_j s(s-1) \ldots (s-j+1) x^{m-j} = 0.$$

But since $s \in (0, 1)$ the product $s(s-1) \ldots (s-j+1)$ never vanishes. Hence the polynomial is identically null if and only if $c_j = 0$ for any j, and we reach a contradiction. This completes the proof of the existence of a function u that satisfies (2.35) and (2.36).

Step 3. Rescaling argument and completion of the proof
By Step 2, for any $m \in \mathbb{N}$ we are able to construct a locally s-harmonic function u such that $u(x) = x^m + \mathcal{O}(x^{m+1})$ near the origin (up to a translation). By considering the blow-up

$$u_\lambda(x) = \frac{u(\lambda x)}{\lambda^m} = x^m + \lambda \mathcal{O}(x^{m+1})$$

we have that for λ small, u_λ is arbitrarily close to the monomial x^m. As stated in Step 1, this concludes the proof of Theorem 2.5.1.

It is worth pointing out that the flexibility of s-harmonic functions given by Theorem 2.5.1 may have concrete consequences. For instance, as a byproduct of Theorem 2.5.1, one has that a biological population with nonlocal dispersive attitudes can better locally adapt to a given distribution of resources (see e.g. Theorem 1.2 in [104]). Namely, nonlocal biological species may efficiently use distant resources and they can fit to the resources available nearby by consuming them (almost) completely, thus making more difficult for a different competing species to come into place.

2.6 A Function with Constant Fractional Laplacian on the Ball

We complete this chapter with the explicit computation of the fractional Laplacian of the function $\mathscr{U}(x) = (1 - |x|^2)^s_+$. In B_1 we have that

$$(-\Delta)^s \mathscr{U}(x) = C(n, s) \frac{\omega_n}{2} B(s, 1 - s),$$

where $C(n, s)$ is introduced in (2.10) and B is the special Beta function (see section 6.2 in [4]). Just to give an idea of how such computation can be obtained, with small modifications respect to [67, 68] we go through the proof of this result. The reader can find the more general result, i.e. for $\mathscr{U}(x) = (1 - |x|^2)^p_+$ for $p > -1$, in the above mentioned [67, 68].

Let us take $u: \mathbb{R} \to \mathbb{R}$ as $u(x) = (1 - |x|^2)^s_+$. We consider the regional fractional Laplacian restricted to $(-1, 1)$

$$\mathscr{L}u(x) := P.V. \int_{-1}^{1} \frac{u(x) - u(y)}{|x - y|^{1+2s}} \, dy.$$

and we compute its value at zero. By symmetry we have that

$$\mathscr{L}u(0) = 2 \lim_{\varepsilon \to 0} \int_{\varepsilon}^{1} \frac{1 - (1 - y^2)^s}{y^{1+2s}} \, dy.$$

Changing the variable $\omega = y^2$ and integrating by parts we get that

$$\mathscr{L}u(0) = 2 \lim_{\varepsilon \to 0} \left(\int_{\varepsilon}^{1} y^{-1-2s} \, dy - \int_{\varepsilon}^{1} (1 - y^2)^s y^{-1-2s} \, dy \right)$$

$$= -\frac{1}{s} + \lim_{\varepsilon \to 0} \left(\frac{\varepsilon^{-2s}}{s} - \int_{\varepsilon^2}^{1} (1 - \omega)^s \omega^{-s-1} \, d\omega \right)$$

$$= -\frac{1}{s} + \lim_{\varepsilon \to 0} \left(\frac{\varepsilon^{-2s} - \varepsilon^{-2s}(1 - \varepsilon^2)^s}{s} + \int_{\varepsilon^2}^{1} \omega^{-s}(1 - \omega)^{s-1} \, d\omega \right).$$

Using the integral representation of the Gamma function (see [4], formula 6.2.1), i.e.

$$B(\alpha, \beta) = \int_0^1 t^{\alpha-1}(1-t)^{\beta-1}\, dt,$$

it yields that

$$\mathcal{L}u(0) = B(1-s, s) - \frac{1}{s}.$$

For $x \in B_1$ we use the change of variables $\omega = \frac{x-y}{1-xy}$. We obtain that

$$\mathcal{L}u(x) = P.V. \int_{-1}^1 \frac{(1-x^2)^s - (1-y^2)^s}{|x-y|^{1+2s}}\, dy$$

$$= (1-x^2)^{-s} P.V. \int_{-1}^1 \frac{(1-\omega x)^{2s-1} - (1-\omega^2)^s(1-\omega x)^{-1}}{|\omega|^{2s+1}}\, d\omega$$

$$= (1-x^2)^{-s} P.V. \left(\int_{-1}^1 \frac{1 - (1-\omega^2)^s}{|\omega|^{2s+1}}\, d\omega + \int_{-1}^1 \frac{(1-\omega x)^{2s-1} - 1}{|\omega|^{2s+1}}\, d\omega \right.$$

$$\left. + \int_{-1}^1 \frac{(1-\omega^2)^s \left(1 - (1-\omega x)^{-1}\right)}{|\omega|^{2s+1}}\, d\omega \right)$$

$$= (1-x^2)^{-s} \left(\mathcal{L}u(0) + J(x) + I(x) \right),$$

$$(2.39)$$

where we have recognized the regional fractional Laplacian and denoted

$$J(x) := P.V. \int_{-1}^1 \frac{(1-\omega x)^{2s-1} - 1}{|\omega|^{2s+1}}\, d\omega \qquad \text{and}$$

$$I(x) := P.V. \int_{-1}^1 \frac{1 - (1-\omega x)^{-1}}{|\omega|^{2s+1}} (1-\omega^2)^s\, d\omega.$$

In $J(x)$ we have that

$$J(x) = P.V. \left(\int_{-1}^1 \frac{(1-\omega x)^{2s-1}}{|\omega|^{2s+1}}\, d\omega - \int_{-1}^1 |\omega|^{-1-2s}\, d\omega \right)$$

$$= \lim_{\varepsilon \to 0} \left(\int_\varepsilon^1 \frac{(1+\omega x)^{2s-1} + (1-\omega x)^{2s-1}}{|\omega|^{2s+1}}\, d\omega - 2\int_\varepsilon^1 \omega^{-1-2s}\, d\omega \right).$$

With the change of variable $t = \dfrac{1}{\omega}$

$$J(x) = \frac{1}{s} + \lim_{\varepsilon \to 0} \left(\int_1^{1/\varepsilon} \left[(t+x)^{2s-1} + (t-x)^{2s-1} \right] dt - \frac{\varepsilon^{-2s}}{s} \right)$$

$$= \frac{1}{s} - \frac{(1+x)^{2s} + (1-x)^{2s}}{2s} + \frac{1}{2s} \lim_{\varepsilon \to 0} \frac{(1+\varepsilon x)^{2s} + (1-\varepsilon x)^{2s} - 2}{\varepsilon^{2s}}$$

$$= \frac{1}{s} - \frac{(1+x)^{2s} + (1-x)^{2s}}{2s}.$$

$$(2.40)$$

To compute $I(x)$, with a Taylor expansion of $(1 - \omega x)^{-1}$ at 0 we have that

$$I(x) = \mathrm{P.V.} \int_{-1}^{1} \frac{-\sum_{k=1}^{\infty}(x\omega)^k}{|\omega|^{2s+1}} (1 - \omega^2)^s \, d\omega.$$

The odd part of the sum vanishes by symmetry, and so

$$I(x) = -2 \lim_{\varepsilon \to 0} \int_{\varepsilon}^{1} \frac{\sum_{k=1}^{\infty}(x\omega)^{2k}}{\omega^{2s+1}} (1 - \omega^2)^s \, d\omega$$

$$= -2 \lim_{\varepsilon \to 0} \sum_{k=1}^{\infty} x^{2k} \int_{\varepsilon}^{1} \omega^{2k-2s-1} (1 - \omega^2)^s \, d\omega.$$

We change the variable $t = \omega^2$ and integrate by parts to obtain

$$I(x) = -\lim_{\varepsilon \to 0} \sum_{k=1}^{\infty} x^{2k} \int_{\varepsilon^2}^{1} t^{k-s-1} (1 - t)^s \, dt,$$

$$= \sum_{k=1}^{\infty} x^{2k} \lim_{\varepsilon \to 0} \left[\frac{\varepsilon^{2k-2s}(1-\varepsilon^2)^s}{k-s} - \frac{s}{k-s} \int_{\varepsilon^2}^{1} t^{k-s}(1-t)^{s-1} \, dt \right].$$

For $k \geq 1$, the limit for ε that goes to zero is null, and using the integral representation of the Beta function, we have that

$$I(x) = \sum_{k=1}^{\infty} x^{2k} \frac{-s}{k-s} B(k+1-s, s).$$

We use the Pochhammer symbol defined as

$$(q)_k = \begin{cases} 1 & \text{for } k = 0, \\ q(q+1)\cdots(q+k-1) & \text{for } k > 0 \end{cases}$$

$$(2.41)$$

and with some manipulations, we get

$$\frac{-s}{k-s}B(k+1-s,s) = \frac{(-s)\Gamma(k+1-s)\Gamma(s)}{(k-s)\Gamma(k+1)}$$

$$= \frac{(-s)\Gamma(k-s)\Gamma(s)}{k!}$$

$$= B(1-s,s)\frac{(-s)_k}{k!}.$$

And so

$$I(x) = B(1-s,s)\sum_{k=1}^{\infty} x^{2k}\frac{(-s)_k}{k!}.$$

By the definition of the hypergeometric function (see e.g. page 211 in [112]) we obtain

$$I(x) = -B(1-s,s) + B(1-s,s)\sum_{k=0}^{\infty}(-s)_k\frac{x^{2k}}{k!}$$

$$= B(1-s,s)\left(F\left(-s,\frac{1}{2},\frac{1}{2},x^2\right)-1\right).$$

Now, by 15.1.8 in [4] we have that

$$F\left(-s,\frac{1}{2},\frac{1}{2},x^2\right) = (1-x^2)^s$$

and therefore

$$I(x) = B(1-s,s)\left((1-x^2)^s - 1\right).$$

Inserting this and (2.40) into (2.39) we obtain

$$\mathscr{L}u(x) = B(1-s,s) - (1-x^2)^{-s}\frac{(1+x)^{2s}+(1-x)^{2s}}{2s}. \qquad (2.42)$$

Now we write the fractional Laplacian of u as

$$\frac{(-\Delta)^s u(x)}{C(1,s)} = \mathscr{L}u(x) + \int_{-\infty}^{-1}\frac{u(x)}{|x-y|^{1+2s}}dy + \int_{1}^{\infty}\frac{u(x)}{|x-y|^{1+2s}}dy$$

$$= \mathscr{L}u(x) + (1-x^2)^s\left(\int_{-\infty}^{-1}|x-y|^{-1-2s}dy + \int_{1}^{\infty}|x-y|^{-1-2s}dy\right)$$

$$= \mathscr{L}u(x) + (1-x^2)^s\frac{(1+x)^{-2s}+(1-x)^{-2s}}{2s}.$$

Inserting (2.42) into the computation, we obtain

$$(-\Delta)^s u(x) = C(1, s) B(1 - s, s).\tag{2.43}$$

To pass to the n-dimensional case, without loss of generality and up to rotations, we consider $x = (0, 0, \ldots, x_n)$ with $x_n \geq 0$. We change into polar coordinates $x - y = th$, with $h \in \partial B_1$ and $t \geq 0$. We have that

$$\frac{(-\Delta)^s \mathscr{U}(x)}{C(n, s)} = P.V. \int_{\mathbb{R}^n} \frac{(1 - |x|^2)^s - (1 - |y|^2)^s}{|x - y|^{n+2s}}\, dy$$

$$= \frac{1}{2} \int_{\partial B_1} \left(P.V. \int_{\mathbb{R}} \frac{(1 - |x|^2)^s - (1 - |x + ht|^2)^s}{|t|^{1+2s}}\, dt \right) d\mathscr{H}^{n-1}(h).\tag{2.44}$$

Changing the variable $t = -|x| h_n + \tau \sqrt{|h_n x|^2 - |x|^2 + 1}$, we notice that

$$1 - |x + ht|^2 = (1 - \tau^2)(1 - |x|^2 + |h_n x|^2)$$

and so

$$P.V. \int_{\mathbb{R}} \frac{(1 - |x|^2)^s - (1 - |x + ht|^2)^s}{|t|^{1+2s}}\, dt$$

$$= P.V. \int_{\mathbb{R}} \frac{(1 - |x|^2)^s - (1 - \tau^2)^s (|h_n x|^2 - |x|^2 + 1)^s}{\left| -|x| h_n + \tau \sqrt{|h_n x|^2 - |x|^2 + 1} \right|^{1+2s}} \sqrt{|h_n x|^2 - |x|^2 + 1}\, d\tau$$

$$= P.V. \int_{\mathbb{R}} \frac{\left(1 - \dfrac{|x|^2 h_n^2}{|h_n x|^2 - |x|^2 + 1} \right)^s - (1 - \tau^2)^s}{\left| \tau - \dfrac{|x| h_n}{\sqrt{|h_n x|^2 - |x|^2 + 1}} \right|^{1+2s}}\, d\tau$$

$$= \frac{(-\Delta)^s u \left(\dfrac{|x| h_n}{\sqrt{|h_n x|^2 - |x|^2 + 1}} \right)}{C(1, s)}$$

$$= B(1 - s, s),$$

where the last equality follows from identity (2.43). Hence from (2.44) we have that

$$(-\Delta)^s \mathscr{U}(x) = C(n, s) B(1 - s, s) \frac{\omega_n}{2}.$$

This concludes the proof of the result.

Chapter 3
Extension Problems

We dedicate this part of the book to the harmonic extension of the fractional Laplacian. We present at first two applications, the water wave model and the Peierls-Nabarro model related to crystal dislocations, making clear how the extension problem appears in these models. We conclude this part by discussing[1] in detail the extension problem via the Fourier transform.

The harmonic extension of the fractional Laplacian in the framework considered here is due to Luis Caffarelli and Luis Silvestre (we refer to [30] for details). We also recall that this extension procedure was obtained by S. A. Molčanov and E. Ostrovskiĭ in [108] by probabilistic methods (roughly speaking "embedding" a long jump random walk in $\mathbb{R}^n$ into a classical random walk in one dimension more, see Fig. 3.1). The idea of this extension procedure is that the nonlocal operator $(-\Delta)^s$ acting on functions defined on $\mathbb{R}^n$ may be reduced to a local operator, acting on functions defined in the higher-dimensional half-space $\mathbb{R}^{n+1}_+ := \mathbb{R}^n \times (0, +\infty)$. Indeed, take $U: \mathbb{R}^{n+1}_+ \to \mathbb{R}$ such that $U(x,0) = u(x)$ in $\mathbb{R}^n$, solution to the equation

$$\operatorname{div}\left(y^{1-2s}\nabla U(x,y)\right) = 0 \quad \text{in} \quad \mathbb{R}^{n+1}_+.$$

Then up to constants one has that

$$-\lim_{y\to 0}\left(y^{1-2s}\partial_y U(x,y)\right) = (-\Delta)^s u(x).$$

---

[1]Though we do not develop this approach here, it is worth mentioning that extended problems arise naturally also from the probabilistic interpretation described in Sect. 1. Roughly speaking, a stochastic process with jumps in $\mathbb{R}^n$ can often be seen as the "trace" of a classical stochastic process in $\mathbb{R}^n \times [0, +\infty)$ (i.e., each time that the classical stochastic process in $\mathbb{R}^n \times [0, +\infty)$ hits $\mathbb{R}^n \times \{0\}$ it induces a jump process over $\mathbb{R}^n$). Similarly, stochastic process with jumps may also be seen as classical processes at discrete, random, time steps.

© Springer International Publishing Switzerland 2016
C. Bucur, E. Valdinoci, *Nonlocal Diffusion and Applications*, Lecture Notes of the Unione Matematica Italiana 20, DOI 10.1007/978-3-319-28739-3_3

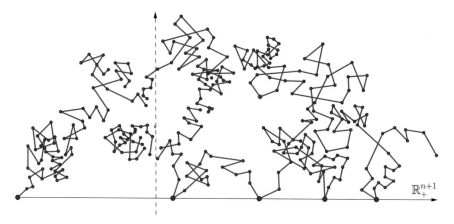

Fig. 3.1 The random walk with jumps in $\mathbb{R}^n$ can be seen as a classical random walk in $\mathbb{R}^{n+1}$

3.1   Water Wave Model

Let us consider the half space $\mathbb{R}^{n+1}_+ = \mathbb{R}^n \times (0, +\infty)$ endowed with the coordinates $x \in \mathbb{R}^n$ and $y \in (0, +\infty)$. We show that the half-Laplacian (namely when $s = 1/2$) arises when looking for a harmonic function in $\mathbb{R}^{n+1}_+$ with given data on $\mathbb{R}^n \times \{y = 0\}$. Thus, let us consider the following local Dirichlet-to-Neumann problem:

$$\begin{cases} \Delta U = 0 & \text{in } \mathbb{R}^{n+1}_+, \\ U(x,0) = u(x) & \text{for } x \in \mathbb{R}^n. \end{cases}$$

The function U is the harmonic extension of u, we write $U = Eu$, and define the operator $\mathcal{L}$ as

$$\mathcal{L}u(x) := -\partial_y U(x, 0). \tag{3.1}$$

We claim that

$$\mathcal{L} = \sqrt{-\Delta_x}, \tag{3.2}$$

in other words

$$\mathcal{L}^2 = -\Delta_x.$$

Indeed, by using the fact that $E(\mathscr{L}u) = -\partial_y U$ (that can be proved, for instance, by using the Poisson kernel representation for the solution), we obtain that

$$
\begin{aligned}
\mathscr{L}^2 u(x) &= \mathscr{L}(\mathscr{L}u)(x) \\
&= -\partial_y E(\mathscr{L}u)(x, 0) \\
&= -\partial_y(-\partial_y U)(x, 0) \\
&= (\partial_{yy} U + \Delta_x U - \Delta_x U)(x, 0) \\
&= \Delta U(x, 0) - \Delta u(x) \\
&= -\Delta u(x),
\end{aligned}
$$

which concludes the proof of (3.2).

One remark in the above calculation lies in the choice of the sign of the square root of the operator. Namely, if we set $\tilde{\mathscr{L}}u(x) := \partial_y U(x, 0)$, the same computation as above would give that $\tilde{\mathscr{L}}^2 = -\Delta$. In a sense, there is no surprise that a quadratic equation offers indeed two possible solutions. But a natural question is how to choose the "right" one.

There are several reasons to justify the sign convention in (3.1). One reason is given by spectral theory, that makes the (fractional) Laplacian a negative definite operator. Let us discuss a purely geometric justification, in the simpler $n = 1$-dimensional case. We wonder how the solution of problem

$$
\begin{cases}
(-\Delta)^s u = 1 & \text{in } (-1, 1), \\
u = 0 & \text{in } \mathbb{R} \setminus (-1, 1).
\end{cases}
\tag{3.3}
$$

should look like in the extended variable y. First of all, by Maximum Principle (recall Theorems 2.3.2 and 2.3.3), we have that u is positive[2] when $x \in (-1, 1)$ (since this is an s-superharmonic function, with zero data outside).

Then the harmonic extension U in $y > 0$ of a function u which is positive in $(-1, 1)$ and vanishes outside $(-1, 1)$ should have the shape of an elastic membrane over the halfplane $\mathbb{R}^2_+$ that is constrained to the graph of u on the trace $\{y = 0\}$. We give a picture of this function U in Fig. 3.2. Notice from the picture that $\partial_y U(x, 0)$ is negative, for any $x \in (-1, 1)$. Since $(-\Delta)^s u(x)$ is positive, we deduce that, to make our picture consistent with the maximum principle, we need to take the sign of $\mathscr{L}$ opposite to that of $\partial_y U(x, 0)$. This gives a geometric justification of (3.1), which is only based on maximum principles (and on "how classical harmonic functions look like").

---

[2]As a matter of fact, the solution of (3.3) is explicit and it is given by $(1 - x^2)^s$, up to dimensional constants. See [68] for a list of functions whose fractional Laplacian can be explicitly computed (unfortunately, differently from the classical cases, explicit computations in the fractional setting are available only for very few functions).

Fig. 3.2 The harmonic
extension

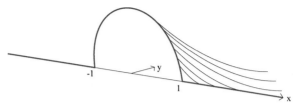

Fig. 3.3 The water waves
model

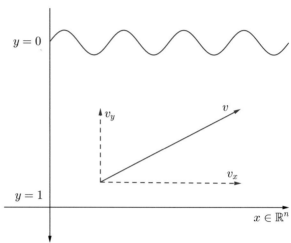

3.1.1 Application to the Water Waves

We show now that the operator $\mathscr{L}$ arises in the theory of water waves of irrotational, incompressible, inviscid fluids in the small amplitude, long wave regime.

Consider a particle moving in the sea, which is, for us, the space $\mathbb{R}^n \times (0, 1)$, where the bottom of the sea is set at level 1 and the surface at level 0 (see Fig. 3.3). The velocity of the particle is $v: \mathbb{R}^n \times (0, 1) \to \mathbb{R}^{n+1}$ and we write $v(x, y) = \big(v_x(x, y), v_y(x, y)\big)$, where $v_x: \mathbb{R}^n \times (0, 1) \to \mathbb{R}^n$ is the horizontal component and $v_y: \mathbb{R}^n \times (0, 1) \to \mathbb{R}$ is the vertical component. We are interested in the vertical velocity of the water at the surface of the sea which we call $u(x)$, namely $u(x) := v_y(x, 0)$. In our model, the water is incompressible, thus div $v = 0$ in $\mathbb{R}^n \times (0, 1)$. Furthermore, on the bottom of sea (since water cannot penetrate into the sand), the velocity has only a non-null horizontal component, hence $v_y(x, 1) = 0$. Also, in our model we assume that there are no vortices: at a mathematical level, this gives that v is irrotational, thus we may write it as the gradient of a function $U: \mathbb{R}^{n+1} \to \mathbb{R}$. We are led to the problem

$$
\begin{cases}
\Delta U = 0 & \text{in } \mathbb{R}^{n+1}_+, \\
\partial_y U(x, 1) = 0 & \text{in } \mathbb{R}^n, \\
U(x, 0) = u(x) & \text{in } \mathbb{R}^n.
\end{cases}
\tag{3.4}
$$

Let $\mathscr{L}$ be, as before, the operator $\mathscr{L}u(x) := -\partial_y U(x,0)$. We solve the problem (3.4) by using the Fourier transform and, up to a normalization factor, we obtain that

$$\mathscr{L}u = \mathscr{F}^{-1}\left(|\xi|\frac{e^{|\xi|} - e^{-|\xi|}}{e^{|\xi|} + e^{-|\xi|}}\hat{u}(\xi)\right).$$

Notice that for large frequencies ξ, this operator is asymptotic to the square root of the Laplacian:

$$\mathscr{L}u \simeq \mathscr{F}^{-1}\left(|\xi|\hat{u}(\xi)\right) = \sqrt{-\Delta}u.$$

The operator $\mathscr{L}$ in the two-dimensional case has an interesting property, that is in analogy to a conjecture of De Giorgi (the forthcoming Sect. 4.2 will give further details about it): more precisely, one considers entire, bounded, smooth, monotone solutions of the equation $\mathscr{L}u = f(u)$ for given f, and proves that the solution only depends on one variable. More precisely:

Theorem 3.1.1 *Let $f \in C^1(\mathbb{R})$ and u be a bounded smooth solution of*

$$\begin{cases} \mathscr{L}u = f(u) & \text{in } \mathbb{R}^2, \\ \partial_{x_2}u > 0 & \text{in } \mathbb{R}^2. \end{cases}$$

Then there exist a direction $\omega \in S^1$ and a function $u_0 \colon \mathbb{R} \to \mathbb{R}$ such that, for any $x \in \mathbb{R}^2$,

$$u(x) = u_0(x \cdot \omega).$$

See Corollary 2 in [52] for a proof of Theorem 3.1.1 and to Theorem 1 in [52] for a more general result (in higher dimension).

3.2 Crystal Dislocation

A crystal is a material whose atoms are displayed in a regular way. Due to some impurities in the material or to an external stress, some atoms may move from their rest positions. The system reacts to small modifications by pushing back towards the equilibrium. Nevertheless, slightly larger modifications may lead to plastic deformations. Indeed, if an atom dislocation is of the order of the periodicity size of the crystal, it can be perfectly compatible with the behavior of the material at a large scale, and it can lead to a permanent modification.

Suitably superposed atom dislocations may also produce macroscopic deformations of the material, and the atom dislocations may be moved by a suitable external
force, which may be more effective if it happens to be compatible with the periodic
structure of the crystal.

These simple considerations may be framed into a mathematical setting, and they
also have concrete applications in many industrial branches (for instance, in the
production of a soda can, in order to change the shape of an aluminium sheet, it
is reasonable to believe that applying the right force to it can be simpler and less
expensive than melting the metal).

It is also quite popular (see e.g. [102]) to describe the atom dislocation motion
in crystals in analogy with the movement of caterpillar (roughly speaking, it is less
expensive for the caterpillar to produce a defect in the alignment of its body and to
dislocate this displacement, rather then rigidly translating his body on the ground).

The mathematical framework of crystal dislocation presented here is related to
the Peierls-Nabarro model, that is a hybrid model in which a discrete dislocation
occurring along a slide line is incorporated in a continuum medium. The total energy
in the Peierls-Nabarro model combines the elastic energy of the material in reaction
to the single dislocations, and the potential energy of the misfit along the glide plane.
The main result is that, at a macroscopic level, dislocations tend to concentrate at
single points, following the natural periodicity of the crystal.

To introduce a mathematical framework for crystal dislocation, first, we "slice"
the crystal with a plane. The mathematical setting will be then, by symmetry
arguments, the half-plane $\mathbb{R}^2_+ = \{(x, y) \in \mathbb{R}^2 \text{ s.t. } y \geq 0\}$ and the glide line will
be the x-axis. In a crystalline structure, the atoms display periodically. Namely,
the atoms on the x-axis have the preference of occupying integer sites. If atoms
move out of their rest position due to a misfit, the material will have an elastic
reaction, trying to restore the crystalline configuration. The tendency is to move
back the atoms to their original positions, or to recreate, by translations, the natural
periodic configuration (see Fig. 3.4). This effect may be modeled by defining
$v^0(x) := v(x, 0)$ to be the discrepancy between the position of the atom x and its

Fig. 3.4 Crystal dislocation

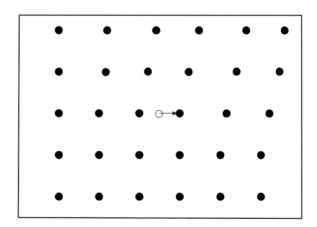

rest position. Then, the misfit energy is

$$\mathcal{M}(v^0) := \int_{\mathbb{R}} W\left(v^0(x)\right) dx, \tag{3.5}$$

where W is a smooth periodic potential, normalized in such a way that $W(u + 1) = W(u)$ for any $u \in \mathbb{R}$ and $0 = W(0) < W(u)$ for any $u \in (0, 1)$. We also assume that the minimum of W is nondegenerate, i.e. $W''(0) > 0$.

We consider the dislocation function $v(x, y)$ on the half-plane $\mathbb{R}^2_+$. The elastic energy of this model is given by

$$\mathcal{E}(v) := \frac{1}{2} \int_{\mathbb{R}^2_+} \left|\nabla v(x, y)\right|^2 dx \, dy. \tag{3.6}$$

The total energy of the system is therefore

$$\mathcal{F}(v) := \mathcal{E}(v) + \mathcal{M}(v^0) = \frac{1}{2} \int_{\mathbb{R}^2_+} \left|\nabla v(x, y)\right|^2 dx \, dy + \int_{\mathbb{R}} W\left(v(x, 0)\right) dx. \tag{3.7}$$

Namely, the total energy of the system is the superposition of the energy in (3.5), which tends to settle all the atoms in their rest position (or in another position equivalent to it from the point of view of the periodic crystal), and the energy in (3.6), which is the elastic energy of the material itself.

Notice that some approximations have been performed in this construction. For instance, the atom dislocation changes the structure of the crystal itself: to write (3.5), one is making the assumption that the dislocations of the single atoms do not destroy the periodicity of the crystal at a large scale, and it is indeed this "permanent" periodic structure that produces the potential W.

Moreover, in (3.6), we are supposing that a "horizontal" atom displacement along the line $\{y = 0\}$ causes a horizontal displacement at $\{y = \epsilon\}$ as well. Of course, in real life, if an atom at $\{y = 0\}$ moves, say, to the right, an atom at level $\{y = \epsilon\}$ is dragged to the right as well, but also slightly downwards towards the slip line $\{y = 0\}$. Thus, in (3.6) we are neglecting this "vertical" displacement. This approximation is nevertheless reasonable since, on the one hand, one expects the vertical displacement to be negligible with respect to the horizontal one and, on the other hand, the vertical periodic structure of the crystal tends to avoid vertical displacements of the atoms outside the periodicity range (from the mathematical point of view, we notice that taking into account vertical displacements would make the dislocation function vectorial, which would produce a system of equations, rather than one single equation for the system).

Also, the initial assumption of slicing the crystal is based on some degree of simplification, since this comes to studying dislocation curves in spaces which are "transversal" to the slice plane.

In any case, we will take these (reasonable, after all) simplifying assumptions for granted, we will study their mathematical consequences and see how the results obtained agree with the physical experience.

To find the Euler-Lagrange equation associated to (3.7), let us consider a perturbation $\phi \in C_0^\infty(\mathbb{R}^2)$, with $\varphi(x) := \phi(x, 0)$ and let v be a minimizer. Then

$$\frac{d}{d\varepsilon} \mathscr{F}(v + \varepsilon\phi)\Big|_{\varepsilon=0} = 0,$$

which gives

$$\int_{\mathbb{R}_+^2} \nabla v \cdot \nabla \phi \, dx \, dy + \int_{\mathbb{R}} W'(v^0)\varphi \, dx = 0.$$

Consider at first the case in which $\operatorname{supp}\phi \cap \partial \mathbb{R}_+^2 = \varnothing$, thus $\varphi = 0$. By the Divergence Theorem we obtain that

$$\int_{\mathbb{R}_+^2} \phi \, \Delta v \, dx \, dy = 0 \quad \text{for any } \phi \in C_0^\infty(\mathbb{R}^2),$$

thus $\Delta v = 0$ in $\mathbb{R}_+^2$. If $\operatorname{supp}\phi \cap \partial \mathbb{R}_+^2 \neq \varnothing$ then we have that

$$0 = \int_{\mathbb{R}_+^2} \operatorname{div}(\phi \nabla v) \, dx \, dy + \int_{\mathbb{R}} W'(v^0)\varphi \, dx$$

$$= \int_{\partial \mathbb{R}_+^2} \phi \frac{\partial v}{\partial v} \, dx + \int_{\mathbb{R}} W'(v^0)\varphi \, dx$$

$$= -\int_{\mathbb{R}} \varphi \frac{\partial v}{\partial y} \, dx + \int_{\mathbb{R}} W'(v^0)\varphi \, dx$$

for an arbitrary $\varphi \in C_0^\infty(\mathbb{R})$ therefore $\dfrac{\partial v}{\partial y}(x, 0) = W'(v^0(x))$ for $x \in \mathbb{R}$. Hence the critical points of $\mathscr{F}$ are solutions of the problem

$$\begin{cases} \Delta v(x, y) = 0 & \text{for } x \in \mathbb{R} \text{ and } y > 0, \\ v(x, 0) = v^0(x) & \text{for } x \in \mathbb{R}, \\ \partial_y v(x, 0) = W'\big(v(x, 0)\big) & \text{for } x \in \mathbb{R} \end{cases}$$

and up to a normalization constant, recalling (3.1) and (3.2), we have that

$$-\sqrt{-\Delta}v(x, 0) = W'\big(v(x, 0)\big), \text{ for any } x \in \mathbb{R}.$$

The corresponding parabolic evolution equation is $\partial_t v(x,0) = -\sqrt{-\Delta} v(x,0) - W'(v(x,0))$.

After this discussion, one is lead to consider the more general case of the fractional Laplacian $(-\Delta)^s$ for any $s \in (0,1)$ (not only the half Laplacian), and the corresponding parabolic equation

$$\partial_t v = -(-\Delta)^s v - W'(v) + \sigma,$$

where σ is a (small) external stress.

If we take the lattice of size ϵ and rescale v and σ as

$$v_\epsilon(t,x) = v\left(\frac{t}{\epsilon^{1+2s}}, \frac{x}{\epsilon}\right) \quad \text{and} \quad \sigma = \epsilon^{2s}\sigma\left(\frac{t}{\epsilon^{1+2s}}, \frac{x}{\epsilon}\right),$$

then the rescaled function satisfies

$$\partial_t v_\epsilon = \frac{1}{\epsilon}\left(-(-\Delta)^s v_\epsilon - \frac{1}{\epsilon^{2s}} W'(v_\epsilon) + \sigma\right) \text{ in } (0,+\infty) \times \mathbb{R} \tag{3.8}$$

with the initial condition

$$v_\epsilon(0,x) = v_\epsilon^0(x) \text{ for } x \in \mathbb{R}.$$

To suitably choose the initial condition v_ϵ^0, we introduce the basic layer[3] solution u, that is, the unique solution of the problem

$$\begin{cases} -(-\Delta)^s u(x) = W'(u) & \text{in } \mathbb{R}, \\ u' > 0 \text{ and } u(-\infty) = 0, u(0) = 1/2, u(+\infty) = 1. \end{cases} \tag{3.9}$$

For the existence of such solution and its main properties see [114] and [25]. Furthermore, the solution decays polynomially at $\pm\infty$ (see [60] and [59]), namely

$$\left| u(x) - H(x) + \frac{1}{2sW''(0)} \frac{x}{|x|^{1+2s}} \right| \leq \frac{C}{|x|^\vartheta} \quad \text{for any } x \in \mathbb{R}^n, \tag{3.10}$$

where $\vartheta > 2s$ and H is the Heaviside step function

$$H(x) = \begin{cases} 1, & x \geq 0 \\ 0, & x < 0. \end{cases}$$

---

[3] As a matter of fact, the solution of (3.9) coincides with the one of a one-dimensional fractional Allen-Cahn equation, that will be discussed in further detail in the forthcoming Sect. 4.1.

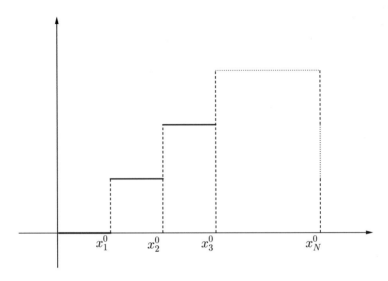

Fig. 3.5 The initial datum when $\varepsilon \to 0$

We take the initial condition of the solution of (3.8) to be the superposition of transitions all occurring with the same orientation, i.e. we set

$$v_\epsilon(x,0) := \frac{\epsilon^{2s}}{W''(0)}\sigma(0,x) + \sum_{i=1}^{N} u\left(\frac{x - x_i^0}{\epsilon}\right), \qquad (3.11)$$

where $x_1^0, \dots, x_N^0$ are N fixed points.

The main result in this setting is that the solution v_ϵ approaches, as $\epsilon \to 0$, the superposition of step functions (Fig. 3.5). The discontinuities of the limit function occur at some points $\left(x_i(t)\right)_{i=1,\dots,N}$ which move accordingly to the following[4] dynamical system

$$\begin{cases} \dot{x}_i = \gamma\left(-\sigma(t,x_i) + \sum_{j\neq i}\dfrac{x_i - x_j}{2s|x_i - x_j|^{2s+1}}\right) & \text{in } (0,+\infty), \\ x_i(0) = x_i^0, \end{cases} \qquad (3.12)$$

where

$$\gamma = \left(\int_{\mathbb{R}}(u')^2\right)^{-1}. \qquad (3.13)$$

More precisely, the main result obtained here is the following.

---

[4] The system of ordinary differential equations in (3.12) has been extensively studied in [80].

Theorem 3.2.1 *There exists a unique viscosity solution of*

$$\begin{cases} \partial_t v_\epsilon = \dfrac{1}{\epsilon}\left(-(-\Delta)^s v_\epsilon - \dfrac{1}{\epsilon^{2s}}W'(v_\epsilon) + \sigma\right) & \text{in } (0,+\infty)\times\mathbb{R}, \\ v_\epsilon(0,x) = \dfrac{\epsilon^{2s}}{W''(0)}\sigma(0,x) + \displaystyle\sum_{i=1}^{N} u\left(\dfrac{x-x_i^0}{\epsilon}\right) & \text{for } x\in\mathbb{R} \end{cases}$$

such that

$$\lim_{\epsilon\to 0} v_\epsilon(t,x) = \sum_{i=1}^{N} H\big(x-x_i(t)\big), \tag{3.14}$$

where $\big(x_i(t)\big)_{i=1,\ldots,N}$ is solution to (3.12).

We refer to [87] for the case $s = \dfrac{1}{2}$, to [60] for the case $s > \dfrac{1}{2}$, and [59] for the case $s < \dfrac{1}{2}$ (in these papers, it is also carefully stated in which sense the limit in (3.14) holds true).

We would like to give now a formal (not rigorous) justification of the ODE system in (3.12) that drives the motion of the transition layers.

Proof (Justification of ODE system (3.12)) We assume for simplicity that the external stress σ is null. We use the notation $\simeq$ to denote the equality up to negligible terms in ϵ. Also, we denote

$$u_i(t,x) := u\left(\frac{x-x_i(t)}{\epsilon}\right)$$

and, with a slight abuse of notation

$$u_i'(t,x) := u'\left(\frac{x-x_i(t)}{\epsilon}\right).$$

By (3.10) we have that the layer solution is approximated by

$$u_i(t,x) \simeq H\left(\frac{x-x_i(t)}{\epsilon}\right) - \frac{\epsilon^{2s}(x-x_i(t))}{2sW''(0)|x-x_i(t)|^{1+2s}}. \tag{3.15}$$

We use the assumption that the solution v_ϵ is well approximated by the sum of N transitions and write

$$v_\epsilon(t,x) \simeq \sum_{i=1}^{N} u_i(t,x) = \sum_{i=1}^{N} u\left(\frac{x-x_i(t)}{\epsilon}\right).$$

For that

$$\partial_t v_\epsilon(t,x) = -\frac{1}{\epsilon} \sum_{i=1}^N u_i'(t,x)\dot{x}_i(t)$$

and, since the basic layer solution u is the solution of (3.9), we have that

$$-(-\varDelta)^s v_\epsilon \simeq -\sum_{i=1}^N (-\varDelta)^s u_i(t,x)$$

$$= -\frac{1}{\epsilon^{2s}} \sum_{i=1}^N (-\varDelta)^s u\left(\frac{x-x_i(t)}{\epsilon}\right)$$

$$= \frac{1}{\epsilon^{2s}} \sum_{i=1}^N W'\left(u\left(\frac{x-x_i(t)}{\epsilon}\right)\right)$$

$$= \frac{1}{\epsilon^{2s}} \sum_{i=1}^N W'\left(u_i(t,x)\right).$$

Now, returning to the parabolic equation (3.8) we have that

$$-\frac{1}{\epsilon} \sum_{i=1}^N u_i'(t,x)\dot{x}_i(t) = \frac{1}{\epsilon^{2s+1}}\left(\sum_{i=1}^N W'\left(u_i(t,x)\right) - W'\left(\sum_{i=1}^N u_i(t,x)\right)\right). \qquad (3.16)$$

Fix an integer k between 1 and N, multiply (3.16) by $u_k'(t,x)$ and integrate over $\mathbb{R}$. We obtain

$$-\frac{1}{\epsilon} \sum_{i=1}^N \dot{x}_i(t) \int_{\mathbb{R}} u_i'(t,x)u_k'(t,x)\,dx$$

$$= \frac{1}{\epsilon^{2s+1}}\left(\sum_{i=1}^N \int_{\mathbb{R}} W'\left(u_i(t,x)\right)u_k'(t,x)\,dx - \int_{\mathbb{R}} W'\left(\sum_{i=1}^N u_i(t,x)\right)u_k'(t,x)\,dx\right).$$

$$(3.17)$$

We compute the left hand side of (3.17). First, we take the kth term of the sum (i.e. we consider the case $i = k$). By using the change of variables

$$y := \frac{x-x_k(t)}{\epsilon} \qquad (3.18)$$

we have that

$$-\frac{1}{\epsilon}\dot{x}_k(t)\int_{\mathbb{R}}(u'_k)^2(t,x)\,dx = -\frac{1}{\epsilon}\dot{x}_k(t)\int_{\mathbb{R}}(u')^2\left(\frac{x-x_k(t)}{\epsilon}\right)dx$$

$$= -\dot{x}_k(t)\int_{\mathbb{R}}(u')^2(y)\,dy \qquad (3.19)$$

$$= -\frac{\dot{x}_k(t)}{\gamma},$$

where γ is defined by (3.13).

Then, we consider the ith term of the sum on the left hand side of (3.17). By performing again the substitution (3.18), we see that this term is

$$-\frac{1}{\epsilon}\dot{x}_i(t)\int_{\mathbb{R}}u'_i(t,x)u'_k(t,x)\,dx = -\frac{1}{\epsilon}\dot{x}_i(t)\int_{\mathbb{R}}u'\left(\frac{x-x_i(t)}{\epsilon}\right)u'\left(\frac{x-x_k(t)}{\epsilon}\right)dx$$

$$= -\dot{x}_i(t)\int_{\mathbb{R}}u'\left(y+\frac{x_k(t)-x_i(t)}{\epsilon}\right)u'(y)\,dy$$

$$\simeq 0,$$

where, for the last equivalence we have used that for ϵ small, $u'\left(y+\dfrac{x_k(t)-x_i(t)}{\epsilon}\right)$ is asymptotic to $u'(\pm\infty)=0$.

We consider the first member on the right hand side of the identity (3.17), and, as before, take the kth term of the sum. We do the substitution (3.18) and have that

$$\frac{1}{\epsilon}\int_{\mathbb{R}}W'\big(u_k(t,x)\big)u'_k(t,x)\,dx = \int_{\mathbb{R}}W'\big(u(y)\big)u'(y)\,dy$$

$$= W\big(u(y)\big)\Big|_{-\infty}^{+\infty}$$

$$= W(1)-W(0)=0$$

by the periodicity of W. Now we use (3.15), the periodicity of W' and we perform a Taylor expansion, noticing that $W'(0)=0$. We see that

$$W'\big(u_i(t,x)\big) \simeq W'\left(H\left(\frac{x-x_i(t)}{\epsilon}\right)-\frac{\epsilon^{2s}(x-x_i(t))}{2sW''(0)|x-x_i(t)|^{1+2s}}\right)$$

$$\simeq W'\left(-\frac{\epsilon^{2s}(x-x_i(t))}{2sW''(0)|x-x_i(t)|^{1+2s}}\right)$$

$$\simeq \frac{-\epsilon^{2s}(x-x_i(t))}{2s|x-x_i(t)|^{1+2s}}.$$

Therefore, the ith term of the sum on the right hand side of the identity (3.17) for $i \neq k$, by using the above approximation and doing one more time the substitution (3.18), for ϵ small becomes

$$\frac{1}{\epsilon}\int_{\mathbb{R}} W'\big(u_i(t,x)\big)u'_k(t,x)\,dx = -\frac{1}{\epsilon}\int_{\mathbb{R}} \frac{\epsilon^{2s}\big(x-x_i(t)\big)}{2s\big|x-x_i(t)\big|^{1+2s}}u'\left(\frac{x-x_k(t)}{\epsilon}\right)dx$$

$$= -\int_{\mathbb{R}} \frac{\epsilon^{2s}\big(\epsilon y+x_k(t)-x_i(t)\big)}{2s\big|\epsilon y+x_k(t)-x_i(t)\big|^{1+2s}}u'(y)\,dy$$

$$\simeq -\frac{\epsilon^{2s}\big(x_k(t)-x_i(t)\big)}{2s\big|x_k(t)-x_i(t)\big|^{1+2s}}\int_{\mathbb{R}} u'(y)\,dy$$

$$= -\frac{\epsilon^{2s}\big(x_k(t)-x_i(t)\big)}{2s\big|x_k(t)-x_i(t)\big|^{1+2s}}.$$

$$(3.20)$$

We also observe that, for ϵ small, the second member on the right hand side of the identity (3.17), by using the change of variables (3.18), reads

$$\frac{1}{\epsilon}\int_{\mathbb{R}} W'\left(\sum_{i=1}^{N} u_i(t,x)\right)u'_k(t,x)\,dx$$

$$= \frac{1}{\epsilon}\int_{\mathbb{R}} W'\left(u_k(t,x)+\sum_{i\neq k} u_i(t,x)\right)u'_k(t,x)\,dx$$

$$= \int_{\mathbb{R}} W'\left(u(y)+\sum_{i\neq k} u\left(y+\frac{x_k(t)-x_i(t)}{\epsilon}\right)\right)u'(y)\,dy.$$

For ϵ small, $u\left(y+\dfrac{x_k(t)-x_i(t)}{\epsilon}\right)$ is asymptotic either to $u(+\infty)=1$ for $x_k > x_i$, or to $u(-\infty)=0$ for $x_k < x_i$. By using the periodicity of W, it follows that

$$\frac{1}{\epsilon}\int_{\mathbb{R}} W'\left(\sum_{i=1}^{N} u_i(t,x)\right)u'_k(t,x)\,dx = \int_{\mathbb{R}} W'\big(u(y)\big)u'(y)\,dy = W(1)-W(0)=0,$$

again by the asymptotic behavior of u. Concluding, by inserting the results (3.19) and (3.20) into (3.17) we get that

$$\frac{\dot{x}_k(t)}{\gamma} = \sum_{i\neq k} \frac{x_k(t)-x_i(t)}{2s\big|x_k(t)-x_i(t)\big|^{1+2s}},$$

which ends the justification of the system (3.12).

We recall that, till now, in Theorem 3.2.1 we considered the initial data as a superposition of transitions all occurring with the same orientation (see (3.11)), i.e. the initial dislocation is a monotone function (all the atoms are initially moved to the right).

Of course, for concrete applications, it is interesting to consider also the case in which the atoms may dislocate in both directions, i.e. the transitions can occur with different orientations (the atoms may be initially displaced to the left or to the right of their equilibrium position).

To model the different orientations of the dislocations, we introduce a parameter $\xi_i \in \{-1, 1\}$ (roughly speaking $\xi_i = 1$ corresponds to a dislocation to the right and $\xi_i = -1$ to a dislocation to the left).

The main result in this case is the following (see [115]):

Theorem 3.2.2 *There exists a viscosity solution of*

$$\begin{cases} \partial_t v_\epsilon = \dfrac{1}{\epsilon}\left(-(-\Delta)^s v_\epsilon - \dfrac{1}{\epsilon^{2s}} W'(v_\epsilon) + \sigma_\epsilon\right) & in\ (0, +\infty) \times \mathbb{R}, \\ v_\epsilon(0, x) = \dfrac{\epsilon^{2s}}{W''(0)}\sigma(0, x) + \displaystyle\sum_{i=1}^N u\left(\xi_i \dfrac{x - x_i^0}{\epsilon}\right) & for\ x \in \mathbb{R} \end{cases}$$

such that

$$\lim_{\epsilon \to 0} v_\epsilon(t, x) = \sum_{i=1}^N H\Big(\xi_i\big(x - x_i(t)\big)\Big),$$

where $\big(x_i(t)\big)_{i=1,\dots,N}$ *is solution to*

$$\begin{cases} \dot{x}_i = \gamma\left(-\xi_i\sigma(t, x_i) + \displaystyle\sum_{j\neq i}\xi_i\xi_j\dfrac{x_i - x_j}{2s|x_i - x_j|^{2s+1}}\right) & in\ (0, +\infty), \\ x_i(0) = x_i^0. \end{cases} \tag{3.21}$$

We observe that Theorem 3.2.2 reduces to Theorem 3.2.1 when $\xi_1 = \cdots = \xi_n = 1$. In fact, the case discussed in Theorem 3.2.2 is richer than the one in Theorem 3.2.1, since, in the case of different initial orientations, collisions can occur, i.e. it may happen that $x_i(T_c) = x_{i+1}(T_c)$ for some $i \in \{1, \dots, N-1\}$ at a collision time T_c.

For instance, in the case $N = 2$, for $\xi_1 = 1$ and $\xi_2 = -1$ (two initial dislocations with different orientations) we have that

$$if\ \sigma \leq 0\ then\ T_c \leq \frac{s\theta_0^{1+2s}}{(2s+1)\gamma},$$

$$if\ \theta_0 < (2s\|\sigma\|_\infty)^{-\frac{1}{2s}}\ then\ T_c \leq \frac{s\theta_0^{1+2s}}{\gamma(1 - 2s\theta_0\|\sigma\|_\infty)},$$

where $\theta_0 := x_2^0 - x_1^0$ is the initial distance between the dislocated atoms. That is, if either the external force has the right sign, or the initial distance is suitably small with respect to the external force, then the dislocation time is finite, and collisions occur in a finite time (on the other hand, when these conditions are violated, there are examples in which collisions do not occur).

This and more general cases of collisions, with precise estimates on the collision times, are discussed in detail in [115].

An interesting feature of the system is that the dislocation function v_ϵ does not annihilate at the collision time. More precisely, in the appropriate scale, we have that v_ϵ at the collision time vanishes outside the collision points, but it still preserves a non-negligible asymptotic contribution exactly at the collision points. A formal statement is the following (see [115]):

Theorem 3.2.3 *Let $N = 2$ and assume that a collision occurs. Let x_c be the collision point, namely $x_c = x_1(T_c) = x_2(T_c)$. Then*

$$\lim_{t \to T_c} \lim_{\varepsilon \to 0} v_\varepsilon(t, x) = 0 \quad \textit{for any} \quad x \neq x_c, \tag{3.22}$$

but

$$\limsup_{\substack{t \to T_c \\ \varepsilon \to 0}} v_\varepsilon(t, x_c) \geq 1. \tag{3.23}$$

Formulas (3.22) and (3.23) describe what happens in the crystal at the collision time. On the one hand, formula (3.22) states that at any point that is not the collision point and at a large scale, the system relaxes at the collision time. On the other hand, formula (3.23) states that the behavior at the collision points at the collision time is quite "singular". Namely, the system does not relax immediately (in the appropriate scale). As a matter of fact, in related numerical simulations (see e.g. [1]) one may notice that the dislocation function may persists after collision and, in higher dimensions, further collisions may change the topology of the dislocation curves.

What happens is that a slightly larger time is needed before the system relaxes exponentially fast: a detailed description of this relaxation phenomenon is presented in [116]. For instance, in the case $N = 2$, the dislocation function decays to zero exponentially fast, slightly after collision, as given by the following result:

Theorem 3.2.4 *Let $N = 2$, $\xi_1 = 1$, $\xi_2 = -1$, and let v_ϵ be the solution given by Theorem 3.2.2, with $\sigma \equiv 0$. Then there exist $\epsilon_0 > 0$, $c > 0$, $T_\epsilon > T_c$ and $\rho_\epsilon > 0$ satisfying*

$$\lim_{\epsilon \to 0} T_\epsilon = T_c$$

$$\textit{and} \quad \lim_{\epsilon \to 0} \varrho_\epsilon = 0$$

such that for any $\epsilon < \epsilon_0$ we have

$$|v_\epsilon(t,x)| \leq \varrho_\epsilon e^{c\frac{T_\epsilon-t}{\epsilon^{2s+1}}}, \quad for\ any\ x \in \mathbb{R}\ and\ t \geq T_\epsilon. \tag{3.24}$$

The estimate in (3.24) states, roughly speaking, that at a suitable time T_ϵ (only slightly bigger than the collision time T_c) the dislocation function gets below a small threshold ρ_ϵ, and later it decays exponentially fast (the constant of this exponential becomes large when ϵ is small).

The reader may compare Theorems 3.2.3 and 3.2.4 and notice that different asymptotics are considered by the two results. A result similar to Theorem 3.2.4 holds for a larger number of dislocated atoms. For instance, in the case of three atoms with alternate dislocations, one has that, slightly after collision, the dislocation function decays exponentially fast to the basic layer solution. More precisely (see again [116]), we have that:

Theorem 3.2.5 *Let $N = 3$, $\xi_1 = \xi_3 = 1$, $\xi_2 = -1$, and let v_ϵ be the solution given by Theorem 3.2.2, with $\sigma \equiv 0$. Then there exist $\epsilon_0 > 0$, $c > 0$, $T_\epsilon^1, T_\epsilon^2 > T_c$ and $\rho_\epsilon > 0$ satisfying*

$$\lim_{\epsilon \to 0} T_\epsilon^1 = \lim_{\epsilon \to 0} T_\epsilon^2 = T_c,$$

$$and \quad \lim_{\epsilon \to 0} \varrho_\epsilon = 0$$

and points $\bar{y}_\epsilon$ and $\bar{z}_\epsilon$ satisfying

$$\lim_{\epsilon \to 0} |\bar{z}_\epsilon - \bar{y}_\epsilon| = 0$$

such that for any $\epsilon < \epsilon_0$ we have

$$v_\epsilon(t,x) \leq u\left(\frac{x-\bar{y}_\epsilon}{\epsilon}\right) + \varrho_\epsilon e^{-\frac{c(t-T_\epsilon^1)}{\epsilon^{2s+1}}}, \quad for\ any\ x \in \mathbb{R}\ and\ t \geq T_\epsilon^1, \tag{3.25}$$

and

$$v_\epsilon(t,x) \geq u\left(\frac{x-\bar{z}_\epsilon}{\epsilon}\right) - \varrho_\epsilon e^{-\frac{c(t-T_\epsilon^2)}{\epsilon^{2s+1}}}, \quad for\ any\ x \in \mathbb{R}\ and\ t \geq T_\epsilon^2, \tag{3.26}$$

where u is the basic layer solution introduced in (3.9).

Roughly speaking, formulas (3.25) and (3.26) say that for times T_ϵ^1, T_ϵ^2 just slightly bigger than the collision time T_c, the dislocation function v_ϵ gets trapped between two basic layer solutions (centered at points $\bar{y}_\epsilon$ and $\bar{z}_\epsilon$), up to a small error. The error gets eventually to zero, exponentially fast in time, and the two basic layer solutions which trap v_ϵ get closer and closer to each other as ϵ goes to zero (that is, the distance between $\bar{y}_\epsilon$ and $\bar{z}_\epsilon$ goes to zero with ϵ).

We refer once more to [116] for a series of figures describing in details the results of Theorems 3.2.4 and 3.2.5. We observe that the results presented in Theorems 3.2.1, 3.2.2, 3.2.3, 3.2.4 and 3.2.5 describe the crystal at different space and time scale. As a matter of fact, the mathematical study of a crystal typically goes from an atomic description (such as a classical discrete model presented by Frenkel-Kontorova and Prandtl-Tomlinson) to a macroscopic scale in which a plastic deformation occurs.

In the theory discussed here, we join this atomic and macroscopic scales by a series of intermediate scales, such as a microscopic scale, in which the Peierls-Nabarro model is introduced, a mesoscopic scale, in which we studied the dynamics of the dislocations (in particular, Theorems 3.2.1 and 3.2.2), in order to obtain at the end a macroscopic theory leading to the relaxation of the model to a permanent deformation (as given in Theorems 3.2.4 and 3.2.5 , while Theorem3.2.3 somehow describes the further intermediate features between these schematic scalings).

3.3 An Approach to the Extension Problem via the Fourier Transform

We will discuss here the extension operator of the fractional Laplacian via the Fourier transform approach (see [30] and [134] for other approaches and further results and also [81], in which a different extension formula is obtained in the framework of the Heisenberg groups).

Some readers may find the details of this part rather technical: if so, she or he can jump directly to Chap. 4 on page 67, without affecting the subsequent reading.

We fix at first a few pieces of notation. We denote points in $\mathbb{R}^{n+1}_+ := \mathbb{R}^n \times (0, +\infty)$ as $X = (x, y)$, with $x \in \mathbb{R}^n$ and $y > 0$. When taking gradients in $\mathbb{R}^{n+1}_+$, we write $\nabla_X = (\nabla_x, \partial_y)$, where ∇_x is the gradient in $\mathbb{R}^n$. Also, in $\mathbb{R}^{n+1}_+$, we will often take the Fourier transform in the variable x only, for fixed $y > 0$. We also set

$$a := 1 - 2s \in (-1, 1).$$

We will consider the fractional Sobolev space $\widehat{H}^s(\mathbb{R}^n)$ defined as the set of functions u that satisfy

$$\|u\|_{L^2(\mathbb{R}^n)} + [\hat{u}]_G < +\infty,$$

where

$$[v]_G := \sqrt{\int_{\mathbb{R}^n} |\xi|^{2s} |v(\xi)|^2 \, d\xi}.$$

For any $u \in W^{1,1}_{loc}((0, +\infty))$, we consider the functional

$$G(u) := \int_0^{+\infty} t^a \left(|u(t)|^2 + |u'(t)|^2 \right) dt. \tag{3.27}$$

By Theorem 4 of [128], we know that the functional G attains its minimum among all the functions $u \in W^{1,1}_{loc}((0, +\infty)) \cap C^0([0, +\infty))$ with $u(0) = 1$. We call g such minimizer and

$$C_\sharp := G(g) = \min_{\substack{u \in W^{1,1}_{loc}((0,+\infty)) \cap C^0([0,+\infty)) \\ u(0)=1}} G(u). \tag{3.28}$$

The main result of this section is the following.

Theorem 3.3.1 *Let $u \in \mathscr{S}(\mathbb{R}^n)$ and let*

$$U(x, y) := \mathscr{F}^{-1}\left(\hat{u}(\xi) \, g(|\xi|y) \right). \tag{3.29}$$

Then

$$\mathrm{div}\,(y^a \nabla U) = 0 \tag{3.30}$$

for any $X = (x, y) \in \mathbb{R}^{n+1}_+$. In addition,

$$-y^a \partial_y U \Big|_{\{y=0\}} = C_\sharp (-\Delta)^s u \tag{3.31}$$

in $\mathbb{R}^n$, both in the sense of distributions and as a pointwise limit.

In order to prove Theorem 3.3.1, we need to make some preliminary computations. At first, let us recall a few useful properties of the minimizer function g of the operator G introduced in (3.27).

We know from formula (4.5) in [128] that

$$0 \le g \le 1, \tag{3.32}$$

and from formula (2.6) in [128] that

$$g' \le 0. \tag{3.33}$$

We also cite formula (4.3) in [128], according to which g is a solution of

$$g''(t) + at^{-1}g'(t) = g(t) \tag{3.34}$$

for any $t > 0$, and formula (4.4) in [128], according to which

$$\lim_{t \to 0^+} t^a g'(t) = -C_\sharp. \tag{3.35}$$

Now, for any $V \in W^{1,1}_{\text{loc}}(\mathbb{R}^{n+1}_+)$ we set

$$[V]_a := \sqrt{\int_{\mathbb{R}^{n+1}_+} y^a |\nabla_X V(X)|^2 \, dX}.$$

Notice that $[V]_a$ is well defined (possibly infinite) on such space. Also, one can compute $[V]_a$ explicitly in the following interesting case:

Lemma 3.1 *Let $\psi \in \mathscr{S}(\mathbb{R}^n)$ and*

$$U(x, y) := \mathscr{F}^{-1}\Big(\psi(\xi) \, g(|\xi|y)\Big). \tag{3.36}$$

Then

$$[U]_a^2 = C_\sharp \, [\psi]_G^2. \tag{3.37}$$

Proof By (3.32), for any fixed $y > 0$, the function $\xi \mapsto \psi(\xi) \, g(|\xi|y)$ belongs to $L^2(\mathbb{R}^n)$, and so we may consider its (inverse) Fourier transform. This says that the definition of U is well posed.

By the inverse Fourier transform definition (2.2), we have that

$$\nabla_x U(x, y) = \nabla_x \int_{\mathbb{R}^n} \psi(\xi) \, g(|\xi|y) \, e^{ix \cdot \xi} \, d\xi$$

$$= \int_{\mathbb{R}^n} i\xi \psi(\xi) \, g(|\xi|y) \, e^{ix \cdot \xi} \, d\xi$$

$$= \mathscr{F}^{-1}\Big(i\xi \psi(\xi) g(|\xi|y)\Big)(x).$$

Thus, by Plancherel Theorem,

$$\int_{\mathbb{R}^n} |\nabla_x U(x, y)|^2 \, dx = \int_{\mathbb{R}^n} \big|\xi \psi(\xi) g(|\xi|y)\big|^2 \, d\xi.$$

Integrating over $y > 0$, we obtain that

$$\int_{\mathbb{R}^{n+1}_+} y^a |\nabla_x U(X)|^2 \, dX = \int_{\mathbb{R}^n} |\xi|^2 |\psi(\xi)|^2 \left[\int_0^{+\infty} y^a |g(|\xi|y)|^2 \, dy \right] d\xi$$

$$= \int_{\mathbb{R}^n} |\xi|^{1-a} |\psi(\xi)|^2 \left[\int_0^{+\infty} t^a |g(t)|^2 \, dt \right] d\xi$$

$$= \int_0^{+\infty} t^a |g(t)|^2 \, dt \cdot \int_{\mathbb{R}^n} |\xi|^{2s} |\psi(\xi)|^2 \, d\xi$$

$$= [\psi]_G^2 \int_0^{+\infty} t^a |g(t)|^2 \, dt. \tag{3.38}$$

Let us now prove that the following identity is well posed

$$\partial_y U(x, y) = \mathscr{F}^{-1}\left(|\xi| \, \psi(\xi) \, g'(|\xi|y) \right). \tag{3.39}$$

For this, we observe that

$$|g'(t)| \le C_\sharp t^{-a}. \tag{3.40}$$

To check this, we define $\gamma(t) := t^a |g'(t)|$. From (3.33) and (3.34), we obtain that

$$\gamma'(t) = -\frac{d}{dt}\left(t^a g'(t) \right) = -t^a \left(g''(t) + at^{-1} g'(t) \right) = -t^a g(t) \le 0.$$

Hence

$$\gamma(t) \le \lim_{\tau \to 0^+} \gamma(\tau) = \lim_{\tau \to 0^+} \tau^a |g'(\tau)| = C_\sharp,$$

where formula (3.35) was used in the last identity, and this establishes (3.40).

From (3.40) we have that $|\xi| \, |\psi(\xi)| \, |g'(|\xi|y)| \le C_\sharp y^{-a} \, |\xi|^{1-a} \, |\psi(\xi)| \in L^2(\mathbb{R}^n)$, and so (3.39) follows.

Therefore, by (3.39) and the Plancherel Theorem,

$$\int_{\mathbb{R}^n} |\partial_y U(x, y)|^2 \, dx = \int_{\mathbb{R}^n} |\xi|^2 \, |\psi(\xi)|^2 \, |g'(|\xi|y)|^2 \, d\xi.$$

Integrating over $y > 0$ we obtain

$$
\begin{aligned}
\int_{\mathbb{R}^{n+1}_+} y^a |\partial_y U(x,y)|^2 \, dx &= \int_{\mathbb{R}^n} |\xi|^2 |\psi(\xi)|^2 \left[\int_0^{+\infty} y^a |g'(|\xi|y)|^2 \, dy \right] d\xi \\
&= \int_{\mathbb{R}^n} |\xi|^{1-a} |\psi(\xi)|^2 \left[\int_0^{+\infty} t^a |g'(t)|^2 dt \right] d\xi \\
&= \int_0^{+\infty} t^a |g'(t)|^2 dt \cdot \int_{\mathbb{R}^n} |\xi|^{2s} |\psi(\xi)|^2 \, d\xi \\
&= [\psi]_G^2 \int_0^{+\infty} t^a |g'(t)|^2 dt.
\end{aligned}
$$

By summing this with (3.38), and recalling (3.28), we obtain the desired result $[U]_a^2 = C_\sharp [\psi]_G^2$. This concludes the proof of the Lemma.

Now, given $u \in L^1_{\text{loc}}(\mathbb{R}^n)$, we consider the space X_u of all the functions $V \in W^{1,1}_{\text{loc}}(\mathbb{R}^{n+1}_+)$ such that, for any $x \in \mathbb{R}^n$, the map $y \mapsto V(x,y)$ is in $C^0([0,+\infty))$, with $V(x,0) = u(x)$ for any $x \in \mathbb{R}^n$. Then the problem of minimizing $[\cdot]_a$ over X_u has a somehow explicit solution.

Lemma 3.2 *Assume that $u \in \mathscr{S}(\mathbb{R}^n)$. Then*

$$
\min_{V \in X_u} [V]_a^2 = [U]_a^2 = C_\sharp [\hat{u}]_G^2, \tag{3.41}
$$

$$
U(x,y) := \mathscr{F}^{-1}\Big(\hat{u}(\xi)\, g(|\xi|y)\Big). \tag{3.42}
$$

Proof We remark that (3.42) is simply (3.36) with $\psi := \hat{u}$, and by Lemma 3.1 we have that

$$
[U]_a^2 = C_\sharp [\hat{u}]_G^2.
$$

Furthermore, we claim that

$$
U \in X_u. \tag{3.43}
$$

In order to prove this, we first observe that

$$
|g(T) - g(t)| \le \frac{C_\sharp |T^{2s} - t^{2s}|}{2s}. \tag{3.44}
$$

To check this, without loss of generality, we may suppose that $T \geq t \geq 0$. Hence, by (3.33) and (3.40),

$$|g(T) - g(t)| \leq \int_t^T |g'(r)| \, dr$$

$$\leq C_\sharp \int_t^T r^{-a} \, dr$$

$$= \frac{C_\sharp (T^{1-a} - t^{1-a})}{1-a},$$

that is (3.44).

Then, by (3.44), for any $y, \tilde{y} \in (0, +\infty)$, we see that

$$\left| g(|\xi| \, y) - g(|\xi| \, \tilde{y}) \right| \leq \frac{C_\sharp |\xi|^{2s} |y^{2s} - \tilde{y}^{2s}|}{2s}.$$

Accordingly,

$$|U(x, y) - U(x, \tilde{y})| = \left| \mathscr{F}^{-1} \left(\hat{u}(\xi) \left(g(|\xi| \, y) - g(|\xi| \, \tilde{y}) \right) \right) \right|$$

$$\leq \int_{\mathbb{R}^n} \left| \hat{u}(\xi) \left(g(|\xi| \, y) - g(|\xi| \, \tilde{y}) \right) \right| d\xi$$

$$\leq \frac{C_\sharp |y^{2s} - \tilde{y}^{2s}|}{2s} \int_{\mathbb{R}^n} |\xi|^{2s} |\hat{u}(\xi)| \, d\xi,$$

and this implies (3.43).

Thanks to (3.43) and (3.37), in order to complete the proof of (3.41), it suffices to show that, for any $V \in X_u$, we have that

$$[V]_a^2 \geq [U]_a^2. \qquad (3.45)$$

To prove this, let us take $V \in X_u$. Without loss of generality, since $[U]_a < +\infty$ thanks to (3.37), we may suppose that $[V]_a < +\infty$. Hence, fixed a.e. $y > 0$, we have that

$$y^a \int_{\mathbb{R}^n} |\nabla_x V(x, y)|^2 \, dx \leq y^a \int_{\mathbb{R}^n} |\nabla_X V(x, y)|^2 \, dx < +\infty,$$

hence the map $x \in |\nabla_x V(x, y)|$ belongs to $L^2(\mathbb{R}^n)$. Therefore, by Plancherel Theorem,

$$\int_{\mathbb{R}^n} |\nabla_x V(x, y)|^2 \, dx = \int_{\mathbb{R}^n} \left| \mathscr{F}(\nabla_x V(x, y))(\xi) \right|^2 d\xi. \tag{3.46}$$

Now by the Fourier transform definition (2.1)

$$\mathscr{F}(\nabla_x V(x, y))(\xi) = \int_{\mathbb{R}^n} \nabla_x V(x, y) \, e^{-ix \cdot \xi} \, dx$$

$$= \int_{\mathbb{R}^n} i\xi \, V(x, y) \, e^{-ix \cdot \xi} \, dx$$

$$= i\xi \, \mathscr{F}(V(x, y))(\xi),$$

hence (3.46) becomes

$$\int_{\mathbb{R}^n} |\nabla_x V(x, y)|^2 \, dx = \int_{\mathbb{R}^n} |\xi|^2 \, |\mathscr{F}(V(x, y))(\xi)|^2 \, d\xi. \tag{3.47}$$

On the other hand

$$\mathscr{F}(\partial_y V(x, y))(\xi) = \partial_y \mathscr{F}(V(x, y))(\xi)$$

and thus, by Plancherel Theorem,

$$\int_{\mathbb{R}^n} |\partial_y V(x, y)|^2 \, dx = \int_{\mathbb{R}^n} |\mathscr{F}(\partial_y V(x, y))(\xi)|^2 \, d\xi = \int_{\mathbb{R}^n} |\partial_y \mathscr{F}(V(x, y))(\xi)|^2 \, d\xi.$$

We sum up this latter result with identity (3.47) and we use the notation $\phi(\xi, y) := \mathscr{F}(V(x, y))(\xi)$ to conclude that

$$\int_{\mathbb{R}^n} |\nabla_X V(x, y)|^2 \, dx = \int_{\mathbb{R}^n} |\xi|^2 \, |\phi(\xi, y)|^2 + |\partial_y \phi(\xi, y)|^2 \, d\xi. \tag{3.48}$$

Accordingly, integrating over $y > 0$, we deduce that

$$[V]_a^2 = \int_{\mathbb{R}^{n+1}_+} y^a \left(|\xi|^2 \, |\phi(\xi, y)|^2 + |\partial_y \phi(\xi, y)|^2 \right) d\xi \, dy. \tag{3.49}$$

Let us first consider the integration over y, for any fixed $\xi \in \mathbb{R}^n \setminus \{0\}$, that we now omit from the notation when this does not generate any confusion. We set $h(y) := \phi(\xi, |\xi|^{-1} y)$. We have that $h'(y) = |\xi|^{-1} \partial_y \phi(\xi, |\xi|^{-1} y)$ and therefore, using the substitution $t = |\xi| y$, we obtain

$$
\begin{aligned}
&\int_0^{+\infty} y^a \left(|\xi|^2 |\phi(\xi, y)|^2 + |\partial_y \phi(\xi, y)|^2 \right) dy \\
&= |\xi|^{1-a} \int_0^{+\infty} t^a \left(|\phi(\xi, |\xi|^{-1} t)|^2 + |\xi|^{-2} |\partial_y \phi(\xi, |\xi|^{-1} t)|^2 \right) dt \\
&= |\xi|^{1-a} \int_0^{+\infty} t^a \left(|h(t)|^2 + |h'(t)|^2 \right) dt \\
&= |\xi|^{2s} G(h).
\end{aligned}
\tag{3.50}
$$

Now, for any $\lambda \in \mathbb{R}$, we show that

$$
\min_{w \in W_{\text{loc}}^{1,1}((0, +\infty)) \cap C^0([0, +\infty))} w(0) = \lambda G(w) = \lambda^2 C_\sharp.
\tag{3.51}
$$

Indeed, when $\lambda = 0$, the trivial function is an allowed competitor and $G(0) = 0$, which gives (3.51) in this case. If, on the other hand, $\lambda \neq 0$, given w as above with $w(0) = \lambda$ we set $w_\lambda(x) := \lambda^{-1} w(x)$. Hence we see that $w_\lambda(0) = 1$ and thus $G(w) = \lambda^2 G(w_\lambda) \leq \lambda^2 G(g) = \lambda^2 C_\sharp$, due to the minimality of g. This proves (3.51). From (3.51) and the fact that

$$
h(0) = \phi(\xi, 0) = \mathscr{F}\big(V(x, 0) \big)(\xi) = \hat{u}(\xi),
$$

we obtain that

$$
G(h) \geq C_\sharp \left| \hat{u}(\xi) \right|^2.
$$

As a consequence, we get from (3.50) that

$$
\int_0^{+\infty} y^a \left(|\xi|^2 |\phi(\xi, y)|^2 + |\partial_y \phi(\xi, y)|^2 \right) dy \geq C_\sharp |\xi|^{2s} \left| \hat{u}(\xi) \right|^2.
$$

Integrating over $\xi \in \mathbb{R}^n \setminus \{0\}$ we obtain that

$$
\int_{\mathbb{R}_+^{n+1}} y^a \left(|\xi|^2 |\phi(\xi, y)|^2 + |\partial_y \phi(\xi, y)|^2 \right) d\xi \, dy \geq C_\sharp [\hat{u}]_G^2.
$$

Hence, by (3.49),

$$
[V]_a^2 \geq C_\sharp [\hat{u}]_G^2,
$$

which proves (3.45), and so (3.41).

We can now prove the main result of this section.

Proof (Proof of Theorem 3.3.1) Formula (3.30) follows from the minimality property in (3.41), by writing that $[U]_a^2 \leq [U + \epsilon \varphi]_a^2$ for any φ smooth and compactly supported inside $\mathbb{R}_+^{n+1}$ and any $\epsilon \in \mathbb{R}$.

Now we take $\varphi \in C_0^\infty(\mathbb{R}^n)$ (notice that its support may now hit $\{y = 0\}$). We define $u_\epsilon := u + \epsilon \varphi$, and U_ϵ as in (3.29), with $\hat{u}$ replaced by $\hat{u}_\epsilon$ (notice that (3.29) is nothing but (3.42)), hence we will be able to exploit Lemma 3.2.

We also set

$$\varphi_*(x, y) := \mathscr{F}^{-1}\Big(\hat{\varphi}(\xi)\, g(|\xi|y)\Big).$$

We observe that

$$\varphi_*(x, 0) = \mathscr{F}^{-1}\Big(\hat{\varphi}(\xi)\, g(0)\Big) = \mathscr{F}^{-1}\Big(\hat{\varphi}(\xi)\Big) = \varphi(x) \tag{3.52}$$

and that

$$U_\epsilon = U + \epsilon \mathscr{F}^{-1}\Big(\hat{\varphi}(\xi)\, g(|\xi|y)\Big) = U + \epsilon \varphi_*.$$

As a consequence

$$[U_\epsilon]_a^2 = [U_\epsilon]_a^2 + 2\epsilon \int_{\mathbb{R}_+^{n+1}} y^a \nabla_X U \cdot \nabla_X \varphi_*\, dX + o(\epsilon).$$

Hence, using (3.30), (3.52) and the Divergence Theorem,

$$[U_\epsilon]_a^2 = [U]_a^2 + 2\epsilon \int_{\mathbb{R}_+^{n+1}} \operatorname{div}\Big(\varphi_*\, y^a \nabla_X U\Big)\, dX + o(\epsilon)$$
$$= [U]_a^2 - 2\epsilon \int_{\mathbb{R}^n \times \{0\}} \varphi\, y^a \partial_y U\, dx + o(\epsilon). \tag{3.53}$$

Moreover, from Plancherel Theorem, and the fact that the image of φ is in the reals,

$$\hat{u}_\epsilon t_G^2 = [\hat{u}]_G + 2\epsilon \int_{\mathbb{R}^n} |\xi|^{2s} \hat{u}(\xi)\, \overline{\hat{\varphi}(\xi)}\, d\xi + o(\epsilon)$$
$$= [\hat{u}]_G + 2\epsilon \int_{\mathbb{R}^n} \mathscr{F}^{-1}\Big(|\xi|^{2s} \hat{u}(\xi)\Big)(x)\, \overline{\varphi(x)}\, dx + o(\epsilon)$$
$$= [\hat{u}]_G + 2\epsilon \int_{\mathbb{R}^n} (-\Delta)^s u(x)\, \varphi(x)\, dx + o(\epsilon).$$

By comparing this with (3.53) and recalling (3.41) we obtain that

$$[U]_a^2 - 2\epsilon \int_{\mathbb{R}^n \times \{0\}} \varphi \, y^a \partial_y U \, dx + o(\epsilon)$$

$$= [U_\epsilon]_a^2$$

$$= C_\sharp [u_\epsilon]_G^2$$

$$= C_\sharp [\hat{u}]_G + 2C_\sharp \epsilon \int_{\mathbb{R}^n} (-\Delta)^s u(x) \, \varphi(x) \, dx + o(\epsilon)$$

$$= [U]_a^2 + 2C_\sharp \epsilon \int_{\mathbb{R}^n} (-\Delta)^s u \, \varphi \, dx + o(\epsilon)$$

and so

$$- \int_{\mathbb{R}^n \times \{0\}} \varphi \, y^a \partial_y U \, dx = C_\sharp \int_{\mathbb{R}^n} (-\Delta)^s u \, \varphi \, dx,$$

for any $\varphi \in C_0^\infty(\mathbb{R}^n)$, that is the distributional formulation of (3.31).
   Furthermore, by (3.29), we have that

$$y^a \partial_y U(x, y) = \mathscr{F}^{-1}\left(|\xi| \, \hat{u}(\xi) \, y^a \, g(|\xi|y) \right) = \mathscr{F}^{-1}\left(|\xi|^{1-a} \, \hat{u}(\xi) \, (|\xi|y)^a \, g(|\xi|y) \right).$$

Hence, by (3.35), we obtain

$$\lim_{y \to 0^+} y^a \partial_y U(x, y) = - C_\sharp \mathscr{F}^{-1}\left(|\xi|^{1-a} \, \hat{u}(\xi) \right)$$

$$= - C_\sharp \mathscr{F}^{-1}\left(|\xi|^{2s} \, \hat{u}(\xi) \right)$$

$$= - (-\Delta)^s u(x),$$

that is the pointwise limit formulation of (3.31). This concludes the proof of
Theorem 3.3.1.

Chapter 4
Nonlocal Phase Transitions

We consider a nonlocal phase transition model, in particular described by the Allen-Cahn equation. A fractional analogue of a conjecture of De Giorgi, that deals with possible one-dimensional symmetry of entire solutions, naturally arises from treating this model, and will be consequently presented. There is a very interesting connection with nonlocal minimal surfaces, that will be studied in Chap. 5.

We introduce briefly the classical case.[1] The Allen-Cahn equation has various applications, for instance, in the study of interfaces (both in gases and solids), in the theory of superconductors and superfluids or in cosmology. We deal here with a two-phase transition model, in which a fluid can reach two pure phases (say 1 and -1) forming an interface of separation. The aim is to describe the pattern and the separation of the two phases.

The formation of the interface is driven by a variational principle. Let $u(x)$ be the function describing the state of the fluid at position x in a bounded region Ω. As a first guess, the phase separation can be modeled via the minimization of the energy

$$\mathcal{E}_0(u) = \int_\Omega W\big(u(x)\big)\, dx,$$

where W is a double-well potential. More precisely, $W: [-1, 1] \to [0, +\infty)$ such that

$$W \in C^2\left([-1, 1]\right), W(\pm 1) = 0, W > 0 \text{ in } (-1, 1),$$
$$W(\pm 1) = 0 \text{ and } W''(\pm 1) > 0. \tag{4.1}$$

---

[1] We would like to thank Alberto Farina who, during a summer-school in Cortona (2014), gave a beautiful introduction on phase transitions in the classical case.

© Springer International Publishing Switzerland 2016
C. Bucur, E. Valdinoci, *Nonlocal Diffusion and Applications*, Lecture Notes
of the Unione Matematica Italiana 20, DOI 10.1007/978-3-319-28739-3_4

The classical example is

$$W(u) := \frac{(u^2 - 1)^2}{4}.$$
(4.2)

On the other hand, the functional in $\mathscr{E}_0$ produces an ambiguous outcome, since any function u that attains only the values ± 1 is a minimizer for the energy. That is, the energy functional in $\mathscr{E}_0$ alone cannot detect any geometric feature of the interface.

To avoid this, one is led to consider an additional energy term that penalizes the formation of unnecessary interfaces. The typical energy functional provided by this procedure has the form

$$\mathscr{E}(u) := \int_\Omega W\big(u(x)\big)\, dx + \frac{\varepsilon^2}{2} \int_\Omega |\nabla u(x)|^2\, dx.$$
(4.3)

In this way, the potential energy that forces the pure phases is compensated by a small term, that is due to the elastic effect of the reaction of the particles. As a curiosity, we point out that in the classical mechanics framework, the analogue of (4.3) is a Lagrangian action of a particle, with $n = 1$, x representing a time coordinate and $u(x)$ the position of the particle at time x. In this framework the term involving the square of the derivative of u has the physical meaning of a kinetic energy. With a slight abuse of notation, we will keep referring to the gradient term in (4.3) as a kinetic energy. Perhaps a more appropriate term would be elastic energy, but in concrete applications also the potential may arise from elastic reactions, therefore the only purpose of these names in our framework is to underline the fact that (4.3) occurs as a superposition of two terms, a potential one, which only depends on u, and one, which will be called kinetic, which only depends on the variation of u (and which, in principle, possesses no real "kinetic" feature).

The energy minimizers will be smooth functions, taking values between -1 and 1, forming layers of interfaces of ε-width. If we send $\varepsilon \to 0$, the transition layer will tend to a minimal surface. To better explain this, consider the energy

$$J(u) = \int \frac{1}{2}|\nabla u|^2 + W(u)\, dx,$$
(4.4)

whose minimizers solve the Allen-Cahn equation

$$-\Delta u + W'(u) = 0.$$
(4.5)

In particular, for the explicit potential in (4.2), Eq. (4.5) reduces (up to normalizations constants) to

$$-\Delta u = u - u^3.$$
(4.6)

In this setting, the behavior of u in large domains reflects into the behavior of the rescaled function $u_\varepsilon(x) = u\left(\frac{x}{\varepsilon}\right)$ in B_1. Namely, the minimizers of J in $B_{1/\varepsilon}$ are the minimizers of J_ε in B_1, where J_ε is the rescaled energy functional

$$J_\varepsilon(u) = \int_{B_1} \frac{\varepsilon}{2}|\nabla u|^2 + \frac{1}{\varepsilon}W(u)\,dx. \qquad (4.7)$$

We notice then that

$$J_\varepsilon(u) \geq \int_{B_1} \sqrt{2W(u)}\,|\nabla u|\,dx$$

which, using the Co-area Formula, gives

$$J_\varepsilon(u) \geq \int_{-1}^{1} \sqrt{2W(t)}\,\mathcal{H}^{n-1}\left(\{u = t\}\right)\,dt.$$

The above formula may suggest that the minimizers of J_ε have the tendency to minimize the $(n-1)$-dimensional measure of their level sets. It turns out that indeed the level sets of the minimizers of J_ε converge to a minimal surface as $\varepsilon \to 0$: for more details see, for instance, [121] and the references therein.

In this setting, a famous De Giorgi conjecture comes into place. In the late 1970s, De Giorgi conjectured that entire, smooth, monotone (in one direction), bounded solutions of (4.6) in the whole of $\mathbb{R}^n$ are necessarily one-dimensional, i.e., there exist $\omega \in S^{n-1}$ and $u_0 : \mathbb{R} \to \mathbb{R}$ such that

$$u(x) = u_0(\omega \cdot x) \quad \text{for any} \quad x \in \mathbb{R}^n.$$

In other words, the conjecture above asks if the level sets of the entire, smooth, monotone (in one direction), bounded solutions are necessarily hyperplanes, at least in dimension $n \leq 8$.

One may wonder why the number eight has a relevance in the problem above. A possible explanation for this is given by the Bernstein Theorem, as we now try to describe.

The Bernstein problem asks on whether or not all minimal graphs (i.e. surfaces that locally minimize the perimeter and that are graphs in a given direction) in $\mathbb{R}^n$ must be necessarily affine. This is indeed true in dimensions n at most eight. On the other hand, in dimension $n \geq 9$ there are global minimal graphs that are not hyperplanes (see e.g. [86]).

The link between the problem of Bernstein and the conjecture of De Giorgi could be suggested by the fact that minimizers approach minimal surfaces in the limit. In a sense, if one is able to prove that the limit interface is a hyperplane and that this rigidity property gets inherited by the level sets of the minimizers u_ε (which lie nearby such limit hyperplane), then, by scaling back, one obtains that the level sets of u are also hyperplanes. Of course, this link between the two problems, as stated here, is only heuristic, and much work is needed to deeply understand the connections between the problem of Bernstein and the conjecture of De Giorgi. We refer to [73] for a more detailed introduction to this topic.

We recall that this conjecture by De Giorgi was proved for $n \leq 3$, see [5, 12, 85]. Also, the case $4 \leq n \leq 8$ with the additional assumption that

$$\lim_{x_n \to \pm\infty} u(x', x_n) = \pm 1, \quad \text{for any} \quad x' \in \mathbb{R}^{n-1} \tag{4.8}$$

was proved in [120].

For $n \geq 9$ a counterexample can be found in [54]. Notice that, if the above limit is uniform (and the De Giorgi conjecture with this additional assumption is known as the Gibbons conjecture), the result extends to all possible n (see for instance [72, 73] for further details).

The goal of the next part of this book is then to discuss an analogue of these questions for the nonlocal case and present related results.

4.1 The Fractional Allen-Cahn Equation

The extension of the Allen-Cahn equation in (4.5) from a local to a nonlocal setting has theoretical interest and concrete applications. Indeed, the study of long range interactions naturally leads to the analysis of phase transitions and interfaces of nonlocal type.

Given an open domain $\Omega \subset \mathbb{R}^n$ and the double well potential W (as in (4.2)), our goal here is to study the fractional Allen-Cahn equation

$$(-\Delta)^s u + W'(u) = 0 \quad \text{in} \quad \Omega,$$

for $s \in (0, 1)$ (when $s = 1$, this equation reduces to (4.5)). The solutions are the critical points of the nonlocal energy

$$\mathscr{E}(u, \Omega) := \int_{\Omega} W\big(u(x)\big) \, dx + \frac{1}{2} \iint_{\mathbb{R}^{2n} \setminus (\Omega^{\mathscr{C}})^2} \frac{|u(x) - u(y)|^2}{|x - y|^{n+2s}} \, dx \, dy, \tag{4.9}$$

up to normalization constants that we omitted for simplicity. The reader can compare (4.9) with (4.3). Namely, in (4.9) the kinetic energy is modified, in order to take into account long range interactions. That is, the new kinetic energy still depends on the variation of the phase parameter. But, in this case, far away changes in phase may influence each other (though the influence is weaker and weaker towards infinity).

Notice that in the nonlocal framework, we prescribe the function on $\Omega^{\mathscr{C}} \times \Omega^{\mathscr{C}}$ and consider the kinetic energy on the remaining regions (see Fig. 4.1). The prescription of values in $\Omega^{\mathscr{C}} \times \Omega^{\mathscr{C}}$ reflects into the fact that the domain of

Fig. 4.1 The kinetic energy

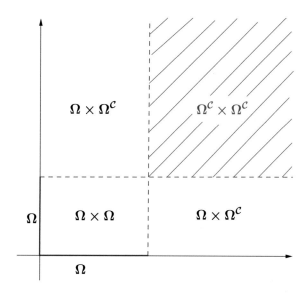

integration of the kinetic integral in (4.9) is $\mathbb{R}^{2n} \setminus (\Omega^{\mathscr{C}})^2$. Indeed, this is perfectly compatible with the local case in (4.3), where the domain of integration of the kinetic term was simply Ω. To see this compatibility, one may think that the domain of integration of the kinetic energy is simply the complement of the set in which the values of the functions are prescribed. In the local case of (4.3), the values are prescribed on $\partial\Omega$, or, one may say, in $\Omega^{\mathscr{C}}$: then the domain of integration of the kinetic energy is the complement of $\Omega^{\mathscr{C}}$, which is simply Ω. In analogy with that, in the nonlocal case of (4.9), the values are prescribed on $\Omega^{\mathscr{C}} \times \Omega^{\mathscr{C}} = (\Omega^{\mathscr{C}})^2$, i.e. outside Ω for both the variables x and y. Then, the kinetic integral is set on the complement of $(\Omega^{\mathscr{C}})^2$, which is indeed $\mathbb{R}^{2n} \setminus (\Omega^{\mathscr{C}})^2$.

Of course, the potential energy has local features, both in the local and in the nonlocal case, since in our model the nonlocality only occurs in the kinetic interaction, therefore the potential integrals are set over Ω both in (4.3) and in (4.9).

For the sake of shortness, given disjoint sets $A, B \subseteq \mathbb{R}^n$ we introduce the notation

$$u(A, B) := \int_A \int_B \frac{|u(x) - u(y)|^2}{|x - y|^{n+2s}} \, dx \, dy,$$

and we write the new kinetic energy in (4.9) as

$$\mathscr{K}(u, \Omega) = \frac{1}{2} u(\Omega, \Omega) + u(\Omega, \Omega^{\mathscr{C}}). \tag{4.10}$$

Let us define the energy minimizers and provide a density estimate for the minimizers.

Definition 4.1.1 The function u is a minimizer for the energy $\mathscr{E}$ in B_R if $\mathscr{E}(u, B_R) \leq \mathscr{E}(v, B_R)$ for any v such that $u = v$ outside B_R.

The energy of the minimizers satisfy the following uniform bound property on large balls.

Theorem 4.1.2 *Let u be a minimizer in B_{R+2} for a large R, say $R \geq 1$. Then*

$$\lim_{R \to +\infty} \frac{1}{R^n} \mathscr{E}(u, B_R) = 0. \tag{4.11}$$

More precisely,

$$\mathscr{E}(u, B_R) \leq \begin{cases} CR^{n-1} & \text{if} \quad s \in \left(\frac{1}{2}, 1\right), \\ CR^{n-1} \log R & \text{if} \quad s = \frac{1}{2}, \\ CR^{n-2s} & \text{if} \quad s \in \left(0, \frac{1}{2}\right). \end{cases}$$

Here, C is a positive constant depending only on n, s and W.

Notice that for $s \in \left(0, \frac{1}{2}\right)$, $R^{n-2s} > R^{n-1}$. These estimates are optimal (we refer to [125] for further details).

Proof We introduce at first some auxiliary functions (see Fig. 4.2). Let

$$\psi(x) := -1 + 2 \min \left\{ (|x| - R - 1)_+, 1 \right\}, \quad v(x) := \min \left\{ u(x), \psi(x) \right\},$$

$$d(x) := \max \left\{ (R + 1 - |x|), 1 \right\}.$$

Then, for $|x - y| \leq d(x)$ we have that

$$|\psi(x) - \psi(y)| \leq \frac{2|x - y|}{d(x)}. \tag{4.12}$$

Indeed, if $|x| \leq R$, then $d(x) = R + 1 - |x|$ and

$$|y| \leq |x - y| + |x| \leq d(x) + |x| \leq R + 1,$$

thus $\psi(x) = \psi(y) = 0$ and the inequality is trivial. Else, if $|x| \geq R$, then $d(x) = 1$, and so the inequality is assured by the Lipschitz continuity of ψ (with 2 as the Lipschitz constant). Also, we prove that we have the following estimates for the function d:

$$\int_{B_{R+2}} d(x)^{-2s}\, dx \leq \begin{cases} CR^{n-1} & \text{if} \quad s \in \left(\frac{1}{2}, 1\right), \\ CR^{n-1} \log R & \text{if} \quad s = \frac{1}{2}, \\ CR^{n-2s} & \text{if} \quad s \in \left(0, \frac{1}{2}\right). \end{cases} \tag{4.13}$$

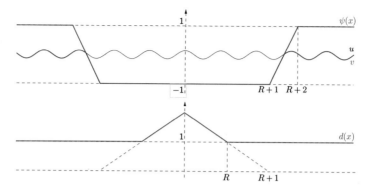

Fig. 4.2 The functions ψ, v and d

To prove this, we observe that in the ring $B_{R+2}\backslash B_R$, we have $d(x) = 1$. Therefore, the contribution to the integral in (4.13) that comes from the ring $B_{R+2} \backslash B_R$ is bounded by the measure of the ring, and so it is of order R^{n-1}, namely

$$\int_{B_{R+2}\backslash B_R} d(x)^{-2s}\, dx = |B_{R+2} \backslash B_R| \leq CR^{n-1}, \tag{4.14}$$

for some $C > 0$. We point out that this order is always negligible with respect to the right hand side of (4.13).

Therefore, to complete the proof of (4.13), it only remains to estimate the contribution to the integral coming from B_R.

For this, we use polar coordinates and perform the change of variables $t = \rho/(R+1)$. In this way, we obtain that

$$\int_{B_R} d(x)^{-2s}\, dx = C \int_0^R \frac{\rho^{n-1}}{(R+1-\rho)^{2s}}\, d\rho$$

$$= C\,(R+1)^{n-2s} \int_0^{1-\frac{1}{R+1}} t^{n-1}(1-t)^{-2s}\, dt$$

$$\leq C\,(R+1)^{n-2s} \int_0^{1-\frac{1}{R+1}} (1-t)^{-2s}\, dt,$$

for some $C > 0$. Now we observe that

$$\int_0^{1-\frac{1}{R+1}} (1-t)^{-2s}\, dt \leq \begin{cases} \int_0^1 (1-t)^{-2s}\, dt = C & \text{if } s \in \left(0, \frac{1}{2}\right), \\ -\log(1-t)\Big|_0^{1-\frac{1}{R+1}} \leq \log R & \text{if } s = \frac{1}{2}, \\ -\frac{(1-t)^{1-2s}}{1-2s}\Big|_0^{1-\frac{1}{R+1}} \leq CR^{2s-1} & \text{if } s \in \left(\frac{1}{2}, 1\right). \end{cases}$$

The latter two formulas and (4.14) imply (4.13).

Now, we define the set

$$A := \{v = \psi\}$$

and notice that $B_{R+1} \subseteq A \subseteq B_{R+2}$. We prove that for any $x \in A$ and any $y \in A^{\mathscr{C}}$

$$|v(x) - v(y)| \leq \max \left\{ |u(x) - u(y)|, |\psi(x) - \psi(y)| \right\}. \tag{4.15}$$

Indeed, for $x \in A$ and $y \in A^{\mathscr{C}}$ we have that

$$v(x) = \psi(x) \leq u(x) \quad \text{and} \quad v(y) = u(y) \leq \psi(y),$$

therefore

$$v(x) - v(y) \leq u(x) - u(y) \quad \text{and} \quad v(y) - v(x) \leq \psi(y) - \psi(x),$$

which establishes (4.15). This leads to

$$v(A, A^{\mathscr{C}}) \leq u(A, A^{\mathscr{C}}) + \psi(A, A^{\mathscr{C}}). \tag{4.16}$$

Notice now that

$$\mathscr{E}(u, B_{R+2}) \leq \mathscr{E}(v, B_{R+2})$$

since u is a minimizer in B_{R+2} and $v = u$ outside B_{R+2}. We have that

$$\mathscr{E}(u, B_{R+2}) = \frac{1}{2} u(B_{R+2}, B_{R+2}) + u(B_{R+2}, B_{R+2}^{\mathscr{C}}) + \int_{B_{R+2}} W(u)\, dx$$

$$= \frac{1}{2} u(A, A) + u(A, A^{\mathscr{C}})$$

$$+ \frac{1}{2} u(B_{R+2} \setminus A, B_{R+2} \setminus A) + u(B_{R+2} \setminus A, B_{R+2}^{\mathscr{C}})$$

$$+ \int_A W(u)\, dx + \int_{B_{R+2} \setminus A} W(u)\, dx.$$

Since u and v coincide on $A^{\mathscr{C}}$, by using the inequality (4.16) we obtain that

$$
\begin{aligned}
0 \;\leq\; & \mathscr{E}(v, B_{R+2}) - \mathscr{E}(u, B_{R+2}) \\
= \;& \frac{1}{2}\, v(A,A) - \frac{1}{2} u(A,A) + v(A,A^{\mathscr{C}}) - u(A,A^{\mathscr{C}}) + \int_A \Big(W(v) - W(u) \Big)\, dx \\
\leq \;& \frac{1}{2}\, v(A,A) - \frac{1}{2} u(A,A) + \psi(A,A^{\mathscr{C}}) + \int_A \Big(W(v) - W(u) \Big)\, dx.
\end{aligned}
$$

Moreover, $v = \psi$ on A and we have that

$$
\frac{1}{2}\, u(A,A) + \int_A W(u)\, dx \leq \frac{1}{2}\, \psi(A,A) + \psi(A,A^{\mathscr{C}}) + \int_A W(\psi)\, dx = \mathscr{E}(\psi,A),
$$

and therefore, since $B_{R+1} \subseteq A \subseteq B_{R+2}$,

$$
\frac{1}{2}\, u(B_{R+1}, B_{R+1}) + \int_{B_{R+1}} W(u)\, dx \leq \mathscr{E}(\psi, B_{R+2}). \tag{4.17}
$$

We estimate now $\mathscr{E}(\psi, B_{R+2})$. For a fixed $x \in B_{R+2}$ we observe that

$$
\begin{aligned}
& \int_{\mathbb{R}^n} \frac{|\psi(x) - \psi(y)|^2}{|x-y|^{n+2s}}\, dy \\
= \;& \int_{|x-y| \leq d(x)} \frac{|\psi(x) - \psi(y)|^2}{|x-y|^{n+2s}}\, dy + \int_{|x-y| \geq d(x)} \frac{|\psi(x) - \psi(y)|^2}{|x-y|^{n+2s}}\, dy \\
\leq \;& C\left(\frac{1}{d(x)^2} \int_{|x-y| \leq d(x)} |x-y|^{-n-2s+2}\, dy + \int_{|x-y| \geq d(x)} |x-y|^{-n-2s}\, dy \right),
\end{aligned}
$$

where we have used (4.12) and the boundedness of ψ. Passing to polar coordinates, we have that

$$
\int_{\mathbb{R}^n} \frac{|\psi(x) - \psi(y)|^2}{|x-y|^{n+2s}}\, dy \leq C\left(\frac{1}{d(x)^2} \int_0^{d(x)} \rho^{-2s+1}\, d\rho + \int_{d(x)}^{\infty} \rho^{-2s-1}\, d\rho \right)
$$

$$
= Cd(x)^{-2s}.
$$

Recalling that $\psi(x) = -1$ on B_{R+1} and $W(-1) = 0$, we obtain that

$$
\begin{aligned}
\mathscr{E}(\psi, B_{R+2}) = \;& \int_{B_{R+2}} \int_{\mathbb{R}^n} \frac{|\psi(x) - \psi(y)|^2}{|x-y|^{n+2s}}\, dy\, dx + \int_{B_{R+2}} W(\psi)\, dx \\
\leq \;& \int_{B_{R+2}} d(x)^{-2s}\, dx + \int_{B_{R+2} \setminus B_{R+1}} W(\psi)\, dx.
\end{aligned}
$$

Therefore, making use of (4.13),

$$
\mathcal{E}(\psi, B_{R+2}) \le
\begin{cases}
CR^{n-1} & \text{if } s \in \left(\frac{1}{2}, 1\right), \\
CR^{n-1} \log R & \text{if } s = \frac{1}{2}, \\
CR^{n-2s} & \text{if } s \in \left(0, \frac{1}{2}\right).
\end{cases}
\tag{4.18}
$$

For what regards the right hand-side of inequality (4.17), we have that

$$
\frac{1}{2} u(B_{R+1}, B_{R+1}) + \int_{B_{R+1}} W(u)\, dx \ge \frac{1}{2} u(B_R, B_R) + u(B_R, B_{R+1} \setminus B_R)
$$
$$
+ \int_{B_R} W(u)\, dx.
\tag{4.19}
$$

We prove now that

$$
u(B_R, B_{R+1}^{\mathscr{C}}) \le \int_{B_{R+2}} d(x)^{-2s}\, dx.
\tag{4.20}
$$

For this, we observe that if $x \in B_R$, then $d(x) = R + 1 - |x|$. So, if $x \in B_R$ and $y \in B_{R+1}^{\mathscr{C}}$, then

$$
|x - y| \ge |y| - |x| \ge R + 1 - |x| = d(x).
$$

Therefore, by changing variables $z = x - y$ and then passing to polar coordinates, we have that

$$
u(B_R, B_{R+1}^{\mathscr{C}}) \le 4 \int_{B_R} dx \int_{B_{d(x)}^{\mathscr{C}}} |z|^{-n-2s}\, dz
$$
$$
\le C \int_{B_R} dx \int_{d(x)}^{\infty} \rho^{-2s-1}\, d\rho
$$
$$
= C \int_{B_R} d(x)^{-2s}\, dx.
$$

This establishes (4.20).

Hence, by (4.13) and (4.20), we have that

$$
u(B_R, B_{R+1}^{\mathscr{C}}) \le \int_{B_{R+2}} d(x)^{-2s}\, dx \le
\begin{cases}
CR^{n-1} & \text{if } s \in \left(\frac{1}{2}, 1\right), \\
CR^{n-1} \log R & \text{if } s = \frac{1}{2}, \\
CR^{n-2s} & \text{if } s \in \left(0, \frac{1}{2}\right).
\end{cases}
\tag{4.21}
$$

We also observe that, by adding $u(B_R, B_{R+1}^{\mathscr{C}})$ to inequality (4.19), we obtain that

$$\frac{1}{2} u(B_{R+1}, B_{R+1}) + \int_{B_{R+1}} W(u) \, dx + u(B_R, B_{R+1}^{\mathscr{C}})$$

$$\geq \frac{1}{2} u(B_R, B_R) + u(B_R, B_{R+1} \setminus B_R) + \int_{B_R} W(u) \, dx + u(B_R, B_{R+1}^{\mathscr{C}})$$

$$= \mathscr{E}(u, B_R).$$

This and (4.17) give that

$$\mathscr{E}(u, B_R) \leq \mathscr{E}(\psi, B_{R+2}) + u(B_R, B_{R+1}^{\mathscr{C}}).$$

Combining this with the estimates in (4.18) and (4.21), we obtain the desired result.

Another type of estimate can be given in terms of the level sets of the minimizers (see Theorem 1.4 in [125]).

Theorem 4.1.3 *Let u be a minimizer of $\mathscr{E}$ in B_R. Then for any $\theta_1, \theta_2 \in (-1, 1)$ such that*

$$u(0) > \theta_1$$

we have that there exist $\overline{R}$ and $C > 0$ such that

$$\left| \{u > \theta_2\} \cap B_R \right| \geq C R^n$$

if $R \geq \overline{R}(\theta_1, \theta_2)$. The constant $C > 0$ depends only on n, s and W and $\overline{R}(\theta_1, \theta_2)$ is a large constant that depends also on θ_1 and θ_2.

The statement of Theorem 4.1.3 says that the level sets of minimizers always occupy a portion of a large ball comparable to the ball itself. In particular, both phases occur in a large ball, and the portion of the ball occupied by each phase is comparable to the one occupied by the other.

Of course, the simplest situation in which two phases split a ball in domains with comparable, and in fact equal, size is when all the level sets are hyperplanes. This question is related to a fractional version of a classical conjecture of De Giorgi and to nonlocal minimal surfaces, that we discuss in the following Sect. 4.2 and Chap. 5.

Let us try now to give some details on the proof of the Theorem 4.1.3 in the particular case in which s is in the range $(0, 1/2)$. The more general proof for all $s \in (0, 1)$ can be found in [125], where one uses some estimates on the Gagliardo norm. In our particular case we will make use of the Sobolev inequality that we introduced in (2.20). The interested reader can see [122] for a more exhaustive explanation of the upcoming proof.

Proof (Proof of Theorem 4.1.3) Let us consider a smooth function w such that $w = 1$ on $B_R^{\mathscr{C}}$ (we will take in sequel w to be a particular barrier for u), and define

$$v(x) := \min\{u(x), w(x)\}.$$

Since $|u| \leq 1$, we have that $v = u$ in $B_R^{\mathscr{C}}$. Calling $D = (\mathbb{R}^n \times \mathbb{R}^n) \setminus \left(B_R^{\mathscr{C}} \times B_R^{\mathscr{C}}\right)$ we have from definition (4.10) that

$$\mathscr{K}(u - v, B_R) + \mathscr{K}(v, B_R) - \mathscr{K}(u, B_R)$$
$$= \frac{1}{2} \iint_D \frac{|(u-v)(x) - (u-v)(y)|^2 + |v(x) - v(y)|^2 - |u(x) - u(y)|^2}{|x-y|^{n+2s}} \, dx \, dy.$$

Using the identity $|a - b|^2 + b^2 - a^2 = 2b(b - a)$ with $a = u(x) - u(y)$ and $b = v(x) - v(y)$ we get

$$\mathscr{K}(u - v, B_R) + \mathscr{K}(v, B_R) - \mathscr{K}(u, B_R)$$
$$= \iint_D \frac{((u-v)(x) - (u-v)(y))\,(v(y) - v(x))}{|x-y|^{n+2s}} \, dx \, dy.$$

Since $u - v = 0$ on $B_R^{\mathscr{C}}$ we can extend the integral to the whole space $\mathbb{R}^n \times \mathbb{R}^n$, hence

$$\mathscr{K}(u - v, B_R) + \mathscr{K}(v, B_R) - \mathscr{K}(u, B_R)$$
$$= \iint_{\mathbb{R}^n \times \mathbb{R}^n} \frac{((u-v)(x) - (u-v)(y))\,(v(y) - v(x))}{|x-y|^{n+2s}} \, dx \, dy.$$

Then by changing variables and using the anti-symmetry of the integrals, we notice that

$$\iint_{B_R \times B_R} \frac{((u-v)(x) - (u-v)(y))\,(v(y) - v(x))}{|x-y|^{n+2s}} \, dx \, dy$$
$$= \iint_{B_R \times B_R} \frac{(u-v)(x)\,(v(y) - v(x))}{|x-y|^{n+2s}} \, dx \, dy$$
$$- \iint_{B_R \times B_R} \frac{(u-v)(y)\,(v(y) - v(x))}{|x-y|^{n+2s}} \, dx \, dy$$
$$= 2 \iint_{B_R \times B_R} \frac{(u-v)(x)\,(v(y) - v(x))}{|x-y|^{n+2s}} \, dx \, dy$$

and

$$\iint_{B_R \times B_R^{\mathscr{C}}} \frac{((u-v)(x) - (u-v)(y)) \, (v(y) - v(x))}{|x-y|^{n+2s}} \, dx \, dy$$

$$+ \iint_{B_R^{\mathscr{C}} \times B_R} \frac{((u-v)(x) - (u-v)(y)) \, (v(y) - v(x))}{|x-y|^{n+2s}} \, dx \, dy$$

$$= \iint_{B_R \times B_R^{\mathscr{C}}} \frac{(u-v)(x) \, (v(y) - v(x))}{|x-y|^{n+2s}} \, dx \, dy$$

$$- \iint_{B_R^{\mathscr{C}} \times B_R} \frac{(u-v)(y) \, (v(y) - v(x))}{|x-y|^{n+2s}} \, dx \, dy$$

$$= 2 \iint_{B_R \times B_R^{\mathscr{C}}} \frac{(u-v)(x) \, (v(y) - v(x))}{|x-y|^{n+2s}} \, dx \, dy.$$

Therefore

$$\mathscr{K}(u-v, B_R) + \mathscr{K}(v, B_R) - \mathscr{K}(u, B_R)$$

$$= 2 \iint_{\mathbb{R}^n \times \mathbb{R}^n} \frac{(u(x) - v(x)) \, (v(y) - v(x))}{|x-y|^{n+2s}} \, dx \, dy$$

$$= 2 \int_{\mathbb{R}^n} (u(x) - v(x)) \left(\int_{\mathbb{R}^n} \frac{v(y) - v(x)}{|x-y|^{n+2s}} \, dy \right) dx$$

$$= 2 \int_{B_R \cap \{u > v = w\}} (u(x) - w(x)) \left(\int_{\mathbb{R}^n} \frac{v(y) - w(x)}{|x-y|^{n+2s}} \, dy \right) dx$$

$$\leq 2 \int_{B_R \cap \{u > v = w\}} (u - w)(x) \left(\int_{\mathbb{R}^n} \frac{w(y) - w(x)}{|x-y|^{n+2s}} \, dy \right) dx$$

$$= 2 \int_{B_R \cap \{u > w\}} (u - w)(x) \, (-(-\Delta)^s w) \, (x) \, dx.$$

Hence

$$\mathscr{K}(u-v, B_R)$$

$$\leq \mathscr{K}(u, B_R) - \mathscr{K}(v, B_R) + 2 \int_{B_R \cap \{u > w\}} (u - w) \, (-(-\Delta)^s w) \, dx.$$

By adding and subtracting the potential energy, we have that

$$\mathcal{K}(u - v, B_R)$$

$$\leq \mathcal{E}(u, B_R) - \mathcal{E}(v, B_R) + \int_{B_R} W(v) - W(u) \, dx$$

$$+ 2 \int_{B_R \cap \{u > w\}} (u - w) \, (-(-\Delta)^s w) \, dx$$

and since u is minimal in B_R,

$$\mathcal{K}(u - v, B_R) \leq \int_{B_R \cap \{u > w = v\}} W(w) - W(u) \, dx$$

$$+ 2 \int_{B_R \cap \{u > w\}} (u - w) \, (-(-\Delta)^s w) \, dx. \tag{4.22}$$

We deduce from the properties in (4.1) of the double-well potential W that there exists a small constant $c > 0$ such that

$$W(t) - W(r) \geq c(1 + r)(t - r) + c(t - r)^2 \quad \text{when } -1 \leq r \leq t \leq -1 + c$$

$$W(r) - W(t) \leq \frac{1 + r}{c} \quad \text{when } -1 \leq r \leq t \leq 1.$$

We fix the arbitrary constants θ_1 and θ_2, take c small as here above. Let then

$$\theta_\star := \min\{\theta_1, \theta_2, -1 + c\}.$$

It follows that

$$\int_{B_R \cap \{u > w\}} W(w) - W(u) \, dx$$

$$= \int_{B_R \cap \{\theta_\star > u > w\}} W(w) - W(u) \, dx + \int_{B_R \cap \{u > \max\{\theta_\star, w\}\}} W(w) - W(u) \, dx$$

$$\leq -c \int_{B_R \cap \{\theta_\star > u > w\}} (1 - w)(u - w) \, dx - c \int_{B_R \cap \{\theta_\star > u > w\}} (u - w)^2 \, dx$$

$$+ \frac{1}{c} \int_{B_R \cap \{u > \max\{\theta_\star, w\}\}} (1 + w) \, dx$$

$$\leq -c \int_{B_R \cap \{\theta_\star > u > w\}} (1 - w)(u - w) \, dx + \frac{1}{c} \int_{B_R \cap \{u > \max\{\theta_\star, w\}\}} (1 + w) \, dx. \tag{4.23}$$

Therefore, in (4.22) we obtain that

$$
\begin{aligned}
\mathscr{K}(u-v, B_R) \leq & -c \int_{B_R \cap \{\theta_* > u > w\}} (1-w)(u-w) \, dx \\
& + \frac{1}{c} \int_{B_R \cap \{u > \max\{\theta_*, w\}\}} (1+w) \, dx \\
& + 2 \int_{B_R \cap \{u > w\}} (u-w)\left(-(-\Delta)^s w\right) \, dx.
\end{aligned}
\tag{4.24}
$$

We introduce now a useful barrier in the next Lemma (we just recall here Lemma 3.1 in [125] – there the reader can find how this barrier is build):

Lemma 4.1 *Given any $\tau \geq 0$ there exists a constant $C > 1$ (possibly depending on n, s and τ) such that: for any $R \geq C$ there exists a rotationally symmetric function $w \in C\left(\mathbb{R}^n, [-1 + CR^{-2s}, 1]\right)$ with $w = 1$ in $B_R^\mathscr{C}$ and such that for any $x \in B_R$ one has that*

$$
\frac{1}{C}(R + 1 - |x|)^{-2s} \leq 1 + w(x) \leq C(R + 1 - |x|)^{-2s} \quad \text{and}
\tag{4.25}
$$

$$
-(-\Delta)^s w(x) \leq \tau(1 + w(x)).
\tag{4.26}
$$

Taking w as the barrier introduced in the above Lemma, thanks to (4.24) and to the estimate in (4.26), we have that

$$
\begin{aligned}
\mathscr{K}(u-v, B_R) \leq & -c \int_{B_R \cap \{\theta_* > u > w\}} (1+w)(u-w) \, dx \\
& + \frac{1}{c} \int_{B_R \cap \{u > \max\{\theta_*, w\}\}} (1+w) \, dx \\
& + 2\tau \int_{B_R \cap \{u > w\}} (u-w)(1+w) \, dx.
\end{aligned}
$$

Let then $\tau = \frac{c}{2}$, and we are left with

$$
\begin{aligned}
\mathscr{K}(u-v, B_R) \leq & \, c \int_{B_R \cap \{u > \max\{\theta_*, w\}\}} (u-w)(1+w) \, dx \\
& + \frac{1}{c} \int_{B_R \cap \{u > \max\{\theta_*, w\}\}} (1+w) \, dx \\
\leq & \, C_1 \int_{B_R \cap \{u > \max\{\theta_*, w\}\}} (1+w) \, dx,
\end{aligned}
$$

with C_1 depending on c (hence on W). Using again Lemma 4.1, in particular the right hand side inequality in (4.25), we have that

$$\mathcal{K}(u - v, B_R) \leq C_1 \cdot C \int_{B_R \cap \{u > \max\{\theta_\star, w\}\}} (R + 1 - |x|)^{-2s}.$$

We set

$$V(R) := |B_R \cap \{u > \theta_\star\}| \tag{4.27}$$

and the Co-Area formula then gives

$$\mathcal{K}(u - v, B_R) \leq C_2 \int_0^R (R + 1 - t)^{-2s} V'(t) \, dt, \tag{4.28}$$

where C_2 possibly depends on n, s, W.

We use now the Sobolev inequality (2.20) for $p = 2$, applied to $u - v$ (recalling that the support of $u - v$ is a subset of B_R) to obtain that

$$\mathcal{K}(u - v, B_R) = \mathcal{K}(u - v, \mathbb{R}^n) = \iint_{\mathbb{R}^n \times \mathbb{R}^n} \frac{|(u - v)(x) - (u - v)(y)|^2}{|x - y|^{n+2s}} \, dx \, dy$$

$$\geq \tilde{C} \|u - v\|_{L^{\frac{2n}{n-2s}}(\mathbb{R}^n)}^2 = \tilde{C} \|u - v\|_{L^{\frac{2n}{n-2s}}(B_R)}^2. \tag{4.29}$$

From (4.25) one has that

$$w(x) \leq C(R + 1 - |x|)^{-2s} - 1.$$

We fix K large enough so as to have $R \geq 2K$ and in B_{R-K}

$$w(x) \leq C(1 + K)^{-2s} - 1 \leq -1 + \frac{1 + \theta_\star}{2}.$$

Therefore in $B_{R-K} \cap \{u > \theta_\star\}$ we have that

$$|u - v| \geq u - w \geq u + 1 - \frac{1 + \theta_\star}{2} \geq \frac{1 + \theta_\star}{2}.$$

Using definition (4.27), this leads to

$$
\|u - v\|^2_{L^{\frac{2n}{n-2s}}(B_R)} = \left(\int_{B_R} |u - v|^{\frac{2n}{n-2s}} dx \right)^{\frac{n-2s}{n}}
$$

$$
\geq \left(\frac{1 + \theta_\star}{2} \right)^{\frac{2n}{n-2s}} \left(\int_{B_{R-K} \cap \{u > \theta_\star\}} dx \right)^{\frac{n-2s}{n}}
$$

$$
\geq C_3 V(R - K)^{\frac{n-2s}{n}}.
$$

In (4.29) we thus have

$$
\mathcal{K}(u - v, B_R) \geq \tilde{C}_3 V(R - K)^{\frac{n-2s}{n}}
$$

and from (4.28) it follows that

$$
C_4 V(R - K)^{\frac{n-2s}{n}} \leq \int_0^R (R + 1 - t)^{-2s} V'(t) \, dt.
$$

Let $R \geq \rho \geq 2K$. Integrating the latter integral from ρ to $\dfrac{3\rho}{2}$ we have that

$$
C_4 \frac{\rho}{2} V(\rho - K)^{\frac{n-2s}{n}} \leq C_4 \int_\rho^{\frac{3\rho}{2}} V(R - K)^{\frac{n-2s}{n}} \, dR
$$

$$
\leq \int_0^{\frac{3\rho}{2}} \left(\int_0^R (R + 1 - t)^{-2s} V'(t) \, dt \right) dR
$$

$$
= \int_0^{\frac{3\rho}{2}} V'(t) \left(\int_0^{\frac{3\rho}{2}} (R + 1 - t)^{-2s} \, dR \right) dt
$$

$$
= \int_0^{\frac{3\rho}{2}} V'(t) \frac{\left(\frac{3\rho}{2} + 1 - t \right)^{1-2s} - 1}{1 - 2s} \, dt.
$$

Since $1 - 2s > 0$, one has for large ρ that $\left(\dfrac{3\rho}{2} + 1 - t \right)^{1-2s} - 1 \leq (2\rho)^{1-2s}$, hence, noticing that the function V is nondecreasing,

$$
\frac{\rho}{2} V(\rho - K)^{\frac{n-2s}{n}} \leq C_5 \rho^{1-2s} \int_0^{2\rho} V'(t) \, dt
$$

$$
\leq C_5 \rho^{1-2s} V(2\rho).
$$

Therefore

$$\rho^{2s}V(\rho-K)^{\frac{n-2s}{n}} \le 2C_5 V(2\rho). \tag{4.30}$$

Now we use an inductive argument as in Lemma 3.2 in [125], that we recall here:

Lemma 4.2 *Let $\sigma, \mu \in (0,\infty)$, $\nu \in (\sigma,\infty)$ and $\gamma, R_0, C \in (1,\infty)$.*
Let $V: (0,\infty) \to (0,\infty)$ be a nondecreasing function. For any $r \in [R_0,\infty)$, let
$\alpha(r) := \min\left\{1, \dfrac{\log V(r)}{\log r}\right\}$. *Suppose that $V(R_0) > \mu$ and*

$$r^{\sigma}\alpha(r)V(r)^{\frac{\nu-\sigma}{\nu}} \le CV(\gamma r),$$

for any $r \in [R_0,\infty)$. Then there exist $c \in (0,1)$ and $R_\star \in [R_0,\infty)$, possibly depending on μ, ν, γ, R_0, C such that

$$V(r) > cr^{\nu},$$

for any $r \in [R_\star,\infty)$.

For R large, one obtains from (4.30) and Lemma 4.2 that

$$V(R) \ge c_0 R^n,$$

for a suitable $c_0 \in (0,1)$. Let now

$$\theta^\star := \max\{\theta_1, \theta_2, -1 + c\}.$$

We have that

$$
\begin{aligned}
|\{u > \theta^\star\} \cap B_R| &+ |\{\theta_\star < u < \theta^\star\} \cap B_R| \\
&= |\{u > \theta_\star\} \cap B_R| \\
&= V(R) \ge c_0 R^n.
\end{aligned}
\tag{4.31}
$$

Moreover, from (4.1.2) we have that for some $\bar{c} > 0$

$$\mathscr{E}(u, B_R) \le \bar{c}R^{n-2s},$$

therefore

$$
\begin{aligned}
\bar{c}R^{n-2s} \ge \mathscr{E}(u, B_R) &\ge \int_{\{\theta_\star < u < \theta^\star\} \cap B_R} W(u)\, dx \\
&\ge \inf_{t \in (\theta_\star, \theta^\star)} W(t)\, |\{\theta_\star < u < \theta^\star\} \cap B_R|.
\end{aligned}
$$

From this and (4.31) we have that

$$c_0 R^n \leq \overline{C} R^{n-2s} + |\{u > \theta^\star\} \cap B_R|,$$

and finally

$$|\{u > \theta^\star\} \cap B_R| \geq C R^n,$$

with C possibly depending on n, s, W. This concludes the proof of Theorem 4.1.3 in the case $s \in (0, 1/2)$.

4.2 A Nonlocal Version of a Conjecture by De Giorgi

In this section we consider the fractional counterpart of the conjecture by De Giorgi that was discussed before in the classical case. Namely, we consider the nonlocal Allen-Cahn equation

$$-(-\Delta)^s u + W(u) = 0 \quad \text{in} \quad \mathbb{R}^n,$$

where W is a double-well potential, and u is smooth, bounded and monotone in one direction, namely $|u| \leq 1$ and $\partial_{x_n} u > 0$. We wonder if it is also true, at least in low dimension, that u is one-dimensional. In this case, the conjecture was initially proved for $n = 2$ and $s = \frac{1}{2}$ in [26]. In the case $n = 2$, for any $s \in (0, 1)$, the result is proved using the harmonic extension of the fractional Laplacian in [25] and [132]. For $n = 3$, the proof can be found in [23] for $s \in \left[\frac{1}{2}, 1\right]$. The conjecture is still open for $n = 3$ and $s \in \left[0, \frac{1}{2}\right]$ and for $n \geq 4$. Also, the Gibbons conjecture (that is the De Giorgi conjecture with the additional condition that limit in (4.8) is uniform) is also true for any $s \in (0, 1)$ and in any dimension n, see [74].

To keep the discussion as simple as possible, we focus here on the case $n = 2$ and any $s \in (0, 1)$, providing an alternative proof that does not make use of the harmonic extension. This part is completely new and not available in the literature. The proof is indeed quite general and it will be further exploited in [42].

We define (as in (4.10)) the total energy of the system to be

$$\mathcal{E}(u, B_R) = \mathcal{K}_R(u) + \int_{B_R} W(u) dx, \tag{4.32}$$

where the kinetic energy is

$$\mathcal{K}_R(u) := \frac{1}{2} \iint_{Q_R} \frac{|u(x) - u(\bar{x})|^2}{|x - \bar{x}|^{n+2s}} dx \, d\bar{x}, \tag{4.33}$$

and $Q_R := \mathbb{R}^{2n} \setminus (B_R^{\mathscr{C}})^2 = (B_R \times B_R) \cup (B_R \times (\mathbb{R}^n \setminus B_R)) \cup ((\mathbb{R}^n \setminus B_R) \times B_R)$. We recall that the kinetic energy can also be written as

$$\mathscr{K}_R(u) = \frac{1}{2}u(B_R, B_R) + u(B_R, B_R^{\mathscr{C}}), \tag{4.34}$$

where for two sets A, B

$$u(A, B) = \int_A \int_B \frac{|u(x) - u(\bar{x})|^2}{|x - \bar{x}|^{n+2s}} \, dx \, d\bar{x}. \tag{4.35}$$

The main result of this section is the following.

Theorem 4.2.1 *Let u be a minimizer of the energy defined in (4.32) in any ball of $\mathbb{R}^2$. Then u is 1-D, i.e. there exist $\omega \in S^1$ and $u_0 : \mathbb{R} \to \mathbb{R}$ such that*

$$u(x) = u_0(\omega \cdot x) \quad \text{for any} \quad x \in \mathbb{R}^2.$$

The proof relies on the following estimate for the kinetic energy, that we prove by employing a domain deformation technique.

Lemma 4.3 *Let $R > 1$, $\varphi \in C_0^\infty(B_1)$. Also, for any $y \in \mathbb{R}^n$, let*

$$\Psi_{R,+}(y) := y + \varphi\left(\frac{y}{R}\right)e_1 \quad and \quad \Psi_{R,-}(y) := y - \varphi\left(\frac{y}{R}\right)e_1. \tag{4.36}$$

Then, for large R, the maps $\Psi_{R,+}$ and $\Psi_{R,-}$ are diffeomorphisms on $\mathbb{R}^n$. Furthermore, if we define $u_{R,\pm}(x) := u(\Psi_{R,\pm}^{-1}(x))$, we have that

$$\mathscr{K}_R(u_{R,+}) + \mathscr{K}_R(u_{R,-}) - 2\mathscr{K}_R(u) \leq \frac{C}{R^2}\mathscr{K}_R(u), \tag{4.37}$$

for some $C > 0$.

Proof First of all, we compute the Jacobian of $\Psi_{R,\pm}$. For this, we write $\Psi_{R,+,i}$ to denote the ith component of the vector $\Psi_{R,+} = (\Psi_{R,+,1}, \cdots, \Psi_{R,+,n})$ and we observe that

$$\frac{\partial \Psi_{R,+,i}(y)}{\partial y_j} = \frac{\partial}{\partial y_j}\left(y_i \pm \varphi\left(\frac{y}{R}\right)\delta_{i1}\right) = \delta_{ij} \pm \frac{1}{R}\partial_j\varphi\left(\frac{y}{R}\right)\delta_{i1}. \tag{4.38}$$

The latter term is bounded by $\mathcal{O}(R^{-1})$, and this proves that $\Psi_{R,\pm}$ is a diffeomorphism if R is large enough.

For further reference, we point out that if $J_{R,\pm}$ is the Jacobian determinant of $\Psi_{R,\pm}$, then the change of variable

$$x := \Psi_{R,\pm}(y), \qquad \bar{x} := \Psi_{R,\pm}(\bar{y}) \tag{4.39}$$

gives that

$$dx\, d\bar{x} = J_{R,\pm}(y)\, J_{R,\pm}(\bar{y})\, dy\, d\bar{y}$$

$$= \left(1 \pm \left(\frac{1}{R}\right)\partial_1\varphi\left(\frac{y}{R}\right) + \mathscr{O}\left(\frac{1}{R^2}\right)\right)\left(1 \pm \frac{1}{R}\partial_1\varphi\left(\frac{\bar{y}}{R}\right) + \mathscr{O}\left(\frac{1}{R^2}\right)\right)dy\, d\bar{y}$$

$$= 1 \pm \frac{1}{R}\partial_1\varphi\left(\frac{y}{R}\right) \pm \frac{1}{R}\partial_1\varphi\left(\frac{\bar{y}}{R}\right) + \mathscr{O}\left(\frac{1}{R^2}\right)dy\, d\bar{y},$$

thanks to (4.38). Therefore

$$\frac{|u_{R,\pm}(x) - u_{R,\pm}(\bar{x})|^2}{|x - \bar{x}|^{n+2s}}\, dx\, d\bar{x}$$

$$= \frac{|u(\Psi_{R,\pm}^{-1}(x)) - u(\Psi_{R,\pm}^{-1}(\bar{x}))|^2}{|\Psi_{R,\pm}^{-1}(x) - \Psi_{R,\pm}^{-1}(\bar{x})|^{n+2s}} \cdot \left(\frac{|x - \bar{x}|^2}{|\Psi_{R,\pm}^{-1}(x) - \Psi_{R,\pm}^{-1}(\bar{x})|^2}\right)^{-\frac{n+2s}{2}} dx\, d\bar{x}$$

$$= \frac{|u(y) - u(\bar{y})|^2}{|y - \bar{y}|^{n+2s}} \cdot \left(\frac{|\Psi_{R,\pm}(y) - \Psi_{R,\pm}(\bar{y})|^2}{|y - \bar{y}|^2}\right)^{-\frac{n+2s}{2}}$$

$$\cdot \left(1 \pm \frac{1}{R}\partial_1\varphi\left(\frac{y}{R}\right) \pm \frac{1}{R}\partial_1\varphi\left(\frac{\bar{y}}{R}\right) + \mathscr{O}\left(\frac{1}{R^2}\right)\right)dy\, d\bar{y}.$$

$$(4.40)$$

Now, for any $y, \bar{y} \in \mathbb{R}^n$ we calculate

$$\left|\Psi_{R,\pm}(y) - \Psi_{R,\pm}(\bar{y})\right|^2$$

$$= \left|(y - \bar{y}) \pm \left(\varphi\left(\frac{y}{R}\right) - \varphi\left(\frac{\bar{y}}{R}\right)\right)e_1\right|^2$$

$$(4.41)$$

$$= |y - \bar{y}|^2 + \left|\varphi\left(\frac{y}{R}\right) - \varphi\left(\frac{\bar{y}}{R}\right)\right|^2 \pm 2\left(\varphi\left(\frac{y}{R}\right) - \varphi\left(\frac{\bar{y}}{R}\right)\right)(y_1 - \bar{y}_1).$$

Notice also that

$$\left|\varphi\left(\frac{y}{R}\right) - \varphi\left(\frac{\bar{y}}{R}\right)\right| \leq \frac{1}{R}\|\varphi\|_{C^1(\mathbb{R}^n)}|y - \bar{y}|,$$

$$(4.42)$$

hence (4.41) becomes

$$\frac{\left|\Psi_{R,\pm}(y) - \Psi_{R,\pm}(\bar{y})\right|^2}{|y - \bar{y}|^2} = 1 + \eta_\pm$$

where

$$\eta_\pm := \frac{\left|\varphi\left(\frac{y}{R}\right) - \varphi\left(\frac{\bar{y}}{R}\right)\right|^2}{|y - \bar{y}|^2} \pm 2\frac{\left(\varphi\left(\frac{y}{R}\right) - \varphi\left(\frac{\bar{y}}{R}\right)\right)(y_1 - \bar{y}_1)}{|y - \bar{y}|^2} = \mathcal{O}\left(\frac{1}{R}\right). \tag{4.43}$$

As a consequence

$$\left(\frac{\left|\Psi_{R,\pm}(y) - \Psi_{R,\pm}(\bar{y})\right|^2}{|y - \bar{y}|^2}\right)^{-\frac{n+2s}{2}} = (1 + \eta_\pm)^{-\frac{n+2s}{2}} = 1 - \frac{n+2s}{2}\eta_\pm + \mathcal{O}(\eta_\pm^2).$$

We plug this information into (4.40) and use (4.43) to obtain

$$\frac{|u_{R,\pm}(x) - u_{R,\pm}(\bar{x})|^2}{|x - \bar{x}|^{n+2s}}\,dx\,d\bar{x}$$

$$= \frac{|u(y) - u(\bar{y})|^2}{|y - \bar{y}|^{n+2s}} \cdot \left(1 - \frac{n+2s}{2}\eta_\pm + \mathcal{O}\left(\frac{1}{R^2}\right)\right)$$

$$\cdot \left(1 \pm \frac{1}{R}\partial_1\varphi\left(\frac{y}{R}\right) \pm \frac{1}{R}\partial_1\varphi\left(\frac{\bar{y}}{R}\right) + \mathcal{O}\left(\frac{1}{R^2}\right)\right)dy\,d\bar{y}$$

$$= \frac{|u(y) - u(\bar{y})|^2}{|y - \bar{y}|^{n+2s}} \cdot \left[1 - \frac{n+2s}{2}\eta_\pm + \left(\pm\frac{1}{R}\partial_1\varphi\left(\frac{y}{R}\right) \pm \frac{1}{R}\partial_1\varphi\left(\frac{\bar{y}}{R}\right)\right)\right.$$

$$\left. + \mathcal{O}\left(\frac{1}{R^2}\right)\right]dy\,d\bar{y}.$$

Using this and the fact that

$$\eta_+ + \eta_- = 2\frac{\left|\varphi\left(\frac{y}{R}\right) - \varphi\left(\frac{\bar{y}}{R}\right)\right|^2}{|y - \bar{y}|^2} = \mathcal{O}\left(\frac{1}{R^2}\right),$$

thanks to (4.42), we obtain

$$\frac{|u_{R,+}(x) - u_{R,+}(\bar{x})|^2}{|x - \bar{x}|^{n+2s}} + \frac{|u_{R,-}(x) - u_{R,-}(\bar{x})|^2}{|x - \bar{x}|^{n+2s}}\,dx\,d\bar{x}$$

$$= \frac{|u(y) - u(\bar{y})|^2}{|y - \bar{y}|^{n+2s}} \cdot \left(2 + \mathcal{O}\left(\frac{1}{R^2}\right)\right)dy\,d\bar{y}.$$

Thus, if we integrate over Q_R we find that

$$\mathscr{K}_R(u_{R,+}) + \mathscr{K}_R(u_{R,+}) = 2\mathscr{K}_R(u) + \iint_{Q_R} \mathscr{O}\left(\frac{1}{R^2}\right) \frac{|u(x) - u(\bar{x})|^2}{|x - \bar{x}|^{n+2s}} \, dx \, d\bar{x}.$$

This establishes (4.37).

Proof (Proof of Theorem 4.2.1) We organize this proof into four steps.
Step 1. A geometrical consideration
   In order to prove that the level sets are flat, it suffices to prove that u is monotone in any direction. Indeed, if u is monotone in any direction, the level set $\{u = 0\}$ is both convex and concave, thus it is flat.
Step 2. Energy estimates
   Let $\varphi \in C_0^\infty(B_1)$ such that $\varphi = 1$ in $B_{1/2}$, and let $e = (1,0)$. We define as in Lemma 4.3

$$\Psi_{R,+}(y) := y + \varphi\left(\frac{y}{R}\right) e \quad \text{and} \quad \Psi_{R,-}(y) := y - \varphi\left(\frac{y}{R}\right) e,$$

which are diffeomorphisms for large R, and the functions $u_{R,\pm}(x) := u(\Psi_{R,+}^{-1}(x))$. Notice that

$$u_{R,+}(y) = u(y) \qquad\qquad \text{for } y \in B_R^{\mathscr{C}} \qquad (4.44)$$

$$u_{R,+}(y) = u(y - e) \qquad\qquad \text{for } y \in B_{R/2}. \qquad (4.45)$$

By computing the potential energy, it is easy to see that

$$\int_{B_R} W(u_{R,+}(x)) \, dx + \int_{B_R} W(u_{R,-}(x)) \, dx - 2 \int_{B_R} W(u(x)) \, dx$$

$$\leq \frac{C}{R^2} \int_{B_R} W(u(x)) \, dx.$$

Using this and (4.37), we obtain the following estimate for the total energy

$$\mathscr{E}(u_{R,+}, B_R) + \mathscr{E}(u_{R,-}, B_R) - 2\mathscr{E}(u, B_R) \leq \frac{C}{R^2}\mathscr{E}(u, B_R). \qquad (4.46)$$

Also, since $u_{R,\pm} = u$ in $B_R^{\mathscr{C}}$, we have that

$$\mathscr{E}(u, B_R) \leq \mathscr{E}(u_{R,-}, B_R).$$

This and (4.46) imply that

$$\mathscr{E}(u_{R,+}, B_R) - \mathscr{E}(u, B_R) \leq \frac{C}{R^2}\mathscr{E}(u, B_R). \qquad (4.47)$$

As a consequence of this estimate and (4.11), it follows that

$$\lim_{R \to +\infty} \left(\mathscr{E}(u_{R,+}, B_R) - \mathscr{E}(u, B_R) \right) = 0. \tag{4.48}$$

Step 3. Monotonicity

We claim that u is monotone. Suppose by contradiction that u is not monotone. That is, up to translation and dilation, we suppose that the value of u at the origin stays above the values of e and $-e$, with $e := (1, 0)$, i.e.

$$u(0) > u(e) \quad \text{and} \quad u(0) > u(-e).$$

Take R to be large enough, say $R > 8$. Let now

$$v_R(x) := \min \{u(x), u_{R,+}(x)\} \quad \text{and} \quad w_R(x) := \max \{u(x), u_{R,+}(x)\}. \tag{4.49}$$

By (4.44) we have that $v_R = w_R = u$ outside B_R. Then, since u is a minimizer in B_R and $w_R = u$ outside B_R, we have that

$$\mathscr{E}(w_R, B_R) \geq \mathscr{E}(u, B_R). \tag{4.50}$$

Moreover, the sum of the energies of the minimum and the maximum is less than or equal to the sum of the original energies: this is obvious in the local case, since equality holds, and in the nonlocal case the proof is based on the inspection of the different integral contributions, see e.g. formula (38) in [114]. So we have that

$$\mathscr{E}(v_R, B_R) + \mathscr{E}(w_R, B_R) \leq \mathscr{E}(u, B_R) + \mathscr{E}(u_{R,+}, B_R)$$

hence, recalling (4.50),

$$\mathscr{E}(v_R, B_R) \leq \mathscr{E}(u_{R,+}, B_R). \tag{4.51}$$

We claim that v_R is not identically neither u, nor $u_{R,+}$. Indeed, since $u(0) = u_{R,+}(e)$ and $u(-e) = u_{R,+}(0)$ we have that

$$v_R(0) = \min \{u(0), u_{R,+}(0)\} = \min \{u(0), u(-e)\}$$
$$= u(-e) = u_{R,+}(0) < u(0) \quad \text{and}$$
$$v_R(e) = \min \{u(e), u_{R,+}(e)\} = \min \{u(e), u(0)\}$$
$$= u(e) < u(0) = u_{R,+}(e).$$

By continuity of u and $u_{R,+}$, we have that

$$v_R = u_{R,+} < u \text{ in a neighborhood of } 0 \quad \text{and}$$
$$v_R = u < u_{R,+} \text{ in a neighborhood of } e. \tag{4.52}$$

We focus our attention on the energy in the smaller ball B_2. We claim that v_R is not minimal for $\mathscr{E}(\cdot, B_2)$. Indeed, if v_R were minimal in B_2, then on B_2 both v_R and u would satisfy the same equation. However, $v_R \leq u$ in $\mathbb{R}^2$ by definition and $v_R = u$ in a neighborhood of e by the second statement in (4.52). The Strong Maximum Principle implies that they coincide everywhere, which contradicts the first line in (4.52).

Hence v_R is not a minimizer in B_2. Let then v_R^* be a minimizer of $\mathscr{E}(\cdot, B_2)$, that agrees with v_R outside the ball B_2, and we define the positive quantity

$$\delta_R := \mathscr{E}(v_R, B_2) - \mathscr{E}(v_R^*, B_2). \tag{4.53}$$

We claim that

$$\text{as } R \text{ goes to infinity, } \delta_R \text{ remains bounded away from zero.} \tag{4.54}$$

To prove this, we assume by contradiction that

$$\lim_{R \to +\infty} \delta_R = 0. \tag{4.55}$$

Consider $\tilde{u}$ to be the translation of u, that is $\tilde{u}(x) := u(x - e)$. Let also

$$m(x) := \min \{u(x), \tilde{u}(x)\}.$$

We notice that in $B_{R/2}$ we have that $\tilde{u}(x) = u_{R,+}(x)$. This and (4.49) give that

$$m = v_R \text{ in } B_{R/2}. \tag{4.56}$$

Also, from (4.52) and (4.56), it follows that m cannot be identically neither u nor $\tilde{u}$, and

$$m < u \text{ in a neighborhood of } 0 \quad \text{and}$$
$$m = u \text{ in a neighborhood of } e. \tag{4.57}$$

Let z be a competitor for m in the ball B_2, that agrees with m outside B_2. We take a cut-off function $\psi \in C_0^\infty(\mathbb{R}^n)$ such that $\psi = 1$ in $B_{R/4}$, $\psi = 0$ in $B_{R/2}^{\mathscr{C}}$. Let

$$z_R(x) := \psi(x)z(x) + \left(1 - \psi(x)\right)v_R(x).$$

Then we have that $z_R = z$ on $B_{R/4}$ and

$$z_R = v_R \text{ on } B_{R/2}^{\mathscr{C}}. \tag{4.58}$$

In addition, by (4.56), we have that $z = m = v_R$ in $B_{R/2} \setminus B_2$. So, it follows that

$$z_R(x) = \psi(x)v_R(x) + (1 - \psi(x))v_R(x) = v_R(x) = z(x) \quad \text{on} \quad B_{R/2} \setminus B_2.$$

This and (4.58) imply that $z_R = v_R$ on $B_2^{\mathscr{C}}$.

We summarize in the next lines these useful identities (see also Fig. 4.3).

$$
\begin{array}{lll}
\text{in } B_2 & u_{R,+} = \tilde{u}, \quad m = v_R, \quad z = z_R \\[4pt]
\text{in } B_{R/2} \setminus B_2 & u_{R,+} = \tilde{u}, \quad v_R^* = v_R = m = z = z_R \\[4pt]
\text{in } B_R \setminus B_{R/2} & v_R^* = v_R = z_R, \quad m = z \\[4pt]
\text{in } B_R^{\mathscr{C}} & u_{R,+} = u = v_R = v_R^* = z_R, \quad m = z.
\end{array}
$$

We compute now

$$\mathscr{E}(m, B_2) - \mathscr{E}(z, B_2)$$
$$= \mathscr{E}(m, B_2) - \mathscr{E}(v_R, B_2) + \mathscr{E}(v_R, B_2) - \mathscr{E}(z_R, B_2) + \mathscr{E}(z_R, B_2) - \mathscr{E}(z, B_2).$$

By the definition of δ_R in (4.53), we have that

$$\mathscr{E}(m, B_2) - \mathscr{E}(z, B_2)$$
$$= \mathscr{E}(m, B_2) - \mathscr{E}(v_R, B_2) + \delta_R + \mathscr{E}(v_R^*, B_2) - \mathscr{E}(z_R, B_2) + \mathscr{E}(z_R, B_2) - \mathscr{E}(z, B_2). \tag{4.59}$$

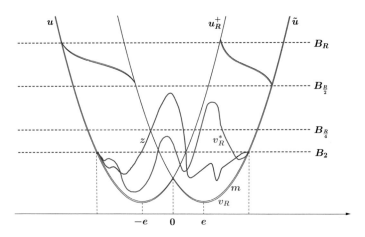

Fig. 4.3 Energy estimates

Using the formula for the kinetic energy given in (4.34) together with (4.35) we have that

$$\mathscr{E}(m, B_2) - \mathscr{E}(v_R, B_2)$$

$$= \frac{1}{2} m(B_2, B_2) + m(B_2, B_2^{\mathscr{C}}) + \int_{B_2} W\big(m(x)\big)\, dx$$

$$- \frac{1}{2} v_R(B_2, B_2) - v_R(B_2, B_2^{\mathscr{C}}) - \int_{B_2} W\big(v_R(x)\big)\, dx.$$

Since $m = v_R$ on $B_{R/2}$ (recall (4.56)), we obtain

$$\mathscr{E}(m, B_2) - \mathscr{E}(v_R, B_2)$$

$$= \int_{B_2} dx \int_{B_{R/2}^{\mathscr{C}}} dy \frac{|m(x) - m(y)|^2 - |m(x) - v_R(y)|^2}{|x - y|^{n+2s}}.$$

Notice now that m and v_R are bounded on $\mathbb{R}^n$ (since so is u). Also, if $x \in B_2$ and $y \in B_{R/2}^{\mathscr{C}}$ we have that $|x - y| \geq |y| - |x| \geq |y|/2$ if R is large. Accordingly,

$$\mathscr{E}(m, B_2) - \mathscr{E}(v_R, B_2) \leq C \int_{B_2} dx \int_{B_{R/2}^{\mathscr{C}}} \frac{1}{|y|^{n+2s}}\, dy \leq C R^{-2s}, \qquad (4.60)$$

up to renaming constants. Similarly, $z_R = z$ on $B_{R/2}$ and we have the same bound

$$\mathscr{E}(z_R, B_2) - \mathscr{E}(z, B_2) \leq C R^{-2s}. \qquad (4.61)$$

Furthermore, since v_R^* is a minimizer for $\mathscr{E}(\cdot, B_2)$ and $v_R^* = z_R$ outside of B_2, we have that

$$\mathscr{E}(v_R^*, B_2) - \mathscr{E}(z_R, B_2) \leq 0.$$

Using this, (4.60) and (4.61) in (4.59), it follows that

$$\mathscr{E}(m, B_2) - \mathscr{E}(z, B_2) \leq C R^{-2s} + \delta_R.$$

Therefore, by sending $R \to +\infty$ and using again (4.55), we obtain that

$$\mathscr{E}(m, B_2) \leq \mathscr{E}(z, B_2). \qquad (4.62)$$

We recall that z can be any competitor for m, that coincides with m outside of B_2. Hence, formula (4.62) means that m is a minimizer for $\mathscr{E}(\cdot, B_2)$. On the other hand, u is a minimizer of the energy in any ball. Then, both u and m satisfy the same equation in B_2. Moreover, they coincide in a neighborhood of e, as stated in the second line

of (4.57). By the Strong Maximum Principle, they have to coincide on B_2, but this
contradicts the first statement of (4.57). The proof of (4.54) is thus complete.

Now, since $v_R^* = v_R$ on $B_2^{\mathscr{C}}$, from definition (4.53) we have that

$$\delta_R = \mathscr{E}(v_R, B_R) - \mathscr{E}(v_R^*, B_R).$$

Also, $\mathscr{E}(v_R^*, B_R) \geq \mathscr{E}(u, B_R)$, thanks to the minimizing property of u. Using these
pieces of information and inequality (4.51), it follows that

$$\delta_R \leq \mathscr{E}(u_{R,+}, B_R) - \mathscr{E}(u, B_R).$$

Now, by sending $R \to +\infty$ and using (4.54), we have that

$$\lim_{R \to +\infty} \mathscr{E}(u_{R,+}, B_R) - \mathscr{E}(u, B_R) > 0,$$

which contradicts (4.48). This implies that indeed u is monotone, and this concludes
the proof of this Step.

Step 4. Conclusions

In Step 3, we have proved that u is monotone, in any given direction e. Then,
Step 1 gives the desired result. This concludes the proof of Theorem 4.2.1.

We remark that the exponent two in the energy estimate (4.37) is related to the
expansions of order two and not to the dimension of the space. Indeed, the energy
estimates hold for any n. However, the two power in the estimate (4.37) allows us to
prove the fractional version of De Giorgi conjecture only in dimension two. In other
words, the proof of Theorem 4.2.1 is not applicable for $n > 2$. One can verify this
by checking the limit in (4.48)

$$\lim_{R \to +\infty} \left(\mathscr{E}(u_{R,+}, B_R) - \mathscr{E}(u, B_R) \right) = 0,$$

which was necessary for the Proof of Theorem 4.2.1 in the case $n = 2$. We know
from Theorem 4.1.2 that

$$\lim_{R \to +\infty} \frac{C}{R^n} \mathscr{E}(u, B_R) = 0.$$

Confronting this result with inequality (4.47)

$$\mathscr{E}(u_{R,+}, B_R) - \mathscr{E}(u, B_R) \leq \frac{C}{R^2} \mathscr{E}(u, B_R),$$

we see that we need to have $n = 2$ in order for the the limit in (4.48) to be zero.

Of course, the one-dimensional symmetry property in Theorem 4.2.1 is inherited
by the spatial homogeneity of the equation, which is translation and rotation
invariant. In the case, for instance, in which the potential also depends on the space

variable, the level sets of the (minimal) solutions may curve, in order to adapt themselves to the spatial inhomogeneity.

Nevertheless, in the case of periodic dependence, it is possible to construct minimal solutions whose level sets are possibly not planar, but still remain at a bounded distance from any fixed hyperplane. As a typical result in this direction, we recall the following one (we refer to [44] for further details on the argument) (Fig. 4.4):

Theorem 4.2.2 *Let $Q_+ > Q_- > 0$ and $Q : \mathbb{R}^n \to [Q_-, Q_+]$. Suppose that $Q(x + k) = Q(x)$ for any $k \in \mathbb{Z}^n$. Let us consider, in any ball B_R, the energy defined by*

$$\mathscr{E}(u, B_R) = \mathscr{K}_R(u) + \frac{1}{4} \int_{B_R} Q(x) \, (1 - u^2)^2 dx,$$

where the kinetic energy $\mathscr{K}_R(u)$ is defined as in (4.33).

Then, there exists a constant $M > 0$, such that, given any $\omega \in \partial B_1$, there exists a minimal solution u_ω of

$$(-\Delta)^s u_\omega(x) = Q(x) \, (u_\omega(x) - u_\omega^3(x)) \qquad \text{for any } x \in \mathbb{R}^n$$

for which the level sets $\{|u_\omega| \leq \frac{9}{10}\}$ are contained in the strip $\{x \in \mathbb{R}^n \text{ s.t. } |\omega \cdot x| \leq M\}$.

Moreover, if ω is rotationally dependent, i.e. if there exists $k_o \in \mathbb{Z}^n$ such that $\omega \cdot k_o = 0$, then u_ω is periodic with respect to ω, i.e.

$$u_\omega(x) = u_\omega(y) \text{ for any } x, y \in \mathbb{R}^n \text{ such that } x - y = k \text{ and } \omega \cdot k = 0.$$

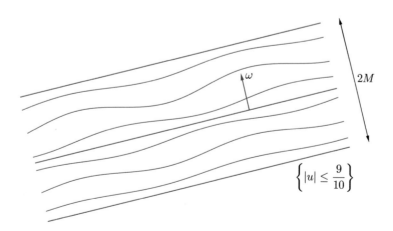

Fig. 4.4 Minimal solutions in periodic medium

Chapter 5
Nonlocal Minimal Surfaces

In this chapter, we introduce nonlocal minimal surfaces and focus on two main results, a Bernstein type result in any dimension and the non-existence of nontrivial s-minimal cones in dimension 2. Moreover, some boundary properties will be discussed at the end of this chapter. For a preliminary introduction to some properties of the nonlocal minimal surfaces, see [135].

Let $\Omega \subset \mathbb{R}^n$ be an open bounded domain, and $E \subset \mathbb{R}^n$ be a measurable set, fixed outside Ω. We will consider for $s \in (0, 1/2)$ minimizers of the H^s norm

$$\|\chi_E\|_{H^s}^2 = \int_{\mathbb{R}^n} \int_{\mathbb{R}^n} \frac{|\chi_E(x) - \chi_E(y)|^2}{|x - y|^{n+2s}} \, dx \, dy$$

$$= 2 \int_{\mathbb{R}^n} \int_{\mathbb{R}^n} \frac{\chi_E(x) \chi_{E^{\mathscr{C}}}(y)}{|x - y|^{n+2s}} \, dx \, dy.$$

Notice that only the interactions between E and $E^{\mathscr{C}}$ contribute to the norm.

In order to define the fractional perimeter of E in Ω, we need to clarify the contribution of Ω to the H^s norm here introduced. Namely, as E is fixed outside Ω, we aim at minimizing the "Ω-contribution" to the norm among all measurable sets that "vary" inside Ω. We consider thus interactions between $E \cap \Omega$ and $E^{\mathscr{C}}$ and between $E \setminus \Omega$ and $\Omega \setminus E$, neglecting the data that is fixed outside Ω and that does not contribute to the minimization of the norm (see Fig. 5.1). We define the interaction $I(A, B)$ of two disjoint subsets of $\mathbb{R}^n$ as

$$I(A, B) := \int_A \int_B \frac{dx \, dy}{|x - y|^{n+2s}}$$

$$= \int_{\mathbb{R}^n} \int_{\mathbb{R}^n} \frac{\chi_A(x) \chi_B(x)}{|x - y|^{n+2s}} \, dx \, dy. \tag{5.1}$$

© Springer International Publishing Switzerland 2016
C. Bucur, E. Valdinoci, *Nonlocal Diffusion and Applications*, Lecture Notes
of the Unione Matematica Italiana 20, DOI 10.1007/978-3-319-28739-3_5

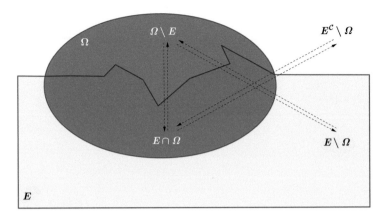

Fig. 5.1 Fractional perimeter

Then (see [28]), one defines the nonlocal s-perimeter functional of E in Ω as

$$\text{Per}_s(E, \Omega) := I(E \cap \Omega, E^{\mathscr{C}}) + I(E \setminus \Omega, \Omega \setminus E). \qquad (5.2)$$

Equivalently, one may write

$$\text{Per}_s(E, \Omega) = I(E \cap \Omega, \Omega \setminus E) + I(E \cap \Omega, E^{\mathscr{C}} \setminus \Omega) + I(E \setminus \Omega, \Omega \setminus E).$$

Definition 5.1 Let Ω be an open domain of $\mathbb{R}^n$. A measurable set $E \subset \mathbb{R}^n$ is s-minimal in Ω if $\text{Per}_s(E, \Omega)$ is finite and if, for any measurable set F such that $E \setminus \Omega = F \setminus \Omega$, we have that

$$\text{Per}_s(E, \Omega) \le \text{Per}_s(F, \Omega).$$

A measurable set is s-minimal in $\mathbb{R}^n$ if it is s-minimal in any ball B_r, where $r > 0$.

When $s \to \frac{1}{2}$, the fractional perimeter Per_s approaches the classical perimeter, see [21]. See also [45] for the precise limit in the class of functions with bounded variations, [34, 35] for a geometric approach towards regularity and [6, 117] for an approach based on Γ-convergence. See also [136] for a different proof and Theorem 2.22 in [103] and the references therein for related discussions. A simple, formal statement, up to renormalizing constants, is the following:

Theorem 5.2 *Let $R > 0$ and E be a set with finite perimeter in B_R. Then*

$$\lim_{s \to \frac{1}{2}} \left(\frac{1}{2} - s \right) Per_s(E, B_r) = Per\,(E, B_r)$$

for almost any $r \in (0, R)$.

The behavior of Per_s as $s \to 0$ is slightly more involved. In principle, the limit as $s \to 0$ of Per_s is, at least locally, related to the Lebesgue measure (see e.g. [106]). Nevertheless, the situation is complicated by the terms coming from infinity, which, as $s \to 0$, become of greater and greater importance. More precisely, it is proved in [58] that, if $\mathrm{Per}_{s_o}(E, \Omega)$ is finite for some $s_o \in (0, 1/2)$, and the limit

$$\beta_E := \lim_{s \to 0} 2s \int_{E \setminus B_1} \frac{dy}{|y|^{n+2s}} \tag{5.3}$$

exists, then

$$\lim_{s \to 0} 2s \, \mathrm{Per}_s(E, \Omega) = (|\partial B_1| - \beta_E) \, |E \cap \Omega| + \beta_E \, |\Omega \setminus E|. \tag{5.4}$$

We remark that, using polar coordinates,

$$0 \leq \beta_E \leq \lim_{s \to 0} 2s \int_{\mathbb{R}^n \setminus B_1} \frac{dy}{|y|^{n+2s}} = \lim_{s \to 0} 2s \, |\partial B_1| \int_1^{+\infty} \rho^{-1-2s} \, d\rho = |\partial B_1|,$$

therefore $\beta_E \in [0, |\partial B_1|]$ plays the role of a convex interpolation parameter in the right hand-side of (5.4) (up to normalization constants).

In this sense, formula (5.4) may be interpreted by saying that, as $s \to 0$, the s-perimeter concentrates itself on two terms that are "localized" in the domain Ω, namely $|E \cap \Omega|$ and $|\Omega \setminus E|$. Nevertheless, the proportion in which these two terms count is given by a "strongly nonlocal" interpolation parameter, namely the quantity β_E in (5.3) which "keeps track" of the behavior of E at infinity.

As a matter of fact, to see how β_E is influenced by the behavior of E at infinity, one can compute β_E for the particular case of a cone. For instance, if $\Sigma \subseteq \partial B_1$, with $\frac{|\Sigma|}{|\partial B_1|} =: b \in [0, 1]$, and E is the cone over Σ (that is $E := \{tp, \; p \in \Sigma, \; t \geq 0\}$), we have that

$$\beta_E = \lim_{s \to 0} 2s \, |\Sigma| \int_1^{+\infty} \rho^{-1-2s} \, d\rho = |\Sigma| = b \, |\partial B_1|,$$

that is β_E gives in this case exactly the opening of the cone.

We also remark that, in general, the limit in (5.3) may not exist, even for smooth sets: indeed, it is possible that the set E "oscillates" wildly at infinity, say from one cone to another one, leading to the non-existence of the limit in (5.3).

Moreover, we point out that the existence of the limit in (5.3) is equivalent to the existence of the limit in (5.4), except in the very special case $|E \cap \Omega| = |\Omega \setminus E|$, in which the limit in (5.4) always exists. That is, the following alternative holds true:

- if $|E \cap \Omega| \neq |\Omega \setminus E|$, then the limit in (5.3) exists if and only if the limit in (5.4) exists,

- if $|E \cap \Omega| = |\Omega \setminus E|$, then the limit in (5.4) always exists (even when the one in (5.3) does not exist), and

$$\lim_{s \to 0} 2s \, \mathrm{Per}_s(E, \Omega) = |\partial B_1| \, |E \cap \Omega| = |\partial B_1| \, |\Omega \setminus E|.$$

The boundaries of s-minimal sets are referred to as *nonlocal minimal surfaces*.

In [28] it is proved that s-minimizers satisfy a suitable integral equation (see in particular Theorem 5.1 in [28]), that is the Euler-Lagrange equation corresponding to the s-perimeter functional Per_s. If E is s-minimal in Ω and ∂E is smooth enough, this Euler-Lagrange equation can be written as

$$\int_{\mathbb{R}^n} \frac{\chi_E(x_0 + y) - \chi_{\mathbb{R}^n \setminus E}(x_0 + y)}{|y|^{n+2s}} \, dy = 0, \tag{5.5}$$

for any $x_0 \in \Omega \cap \partial E$.

Therefore, in analogy with the case of the classical minimal surfaces, which have zero mean curvature, one defines the *nonlocal mean curvature* of E at $x_0 \in \partial E$ as

$$H_E^s(x_0) := \int_{\mathbb{R}^n} \frac{\chi_E(y) - \chi_{E^{\mathscr{C}}}(y)}{|y - x_0|^{n+2s}} \, dy. \tag{5.6}$$

In this way, Eq. (5.5) can be written as $H_E^s = 0$ along ∂E.

It is also suggestive to think that the function $\tilde{\chi}_E := \chi_E - \chi_{E^{\mathscr{C}}}$ averages out to zero at the points on ∂E, if ∂E is smooth enough, since at these points the local contribution of E compensates the one of $E^{\mathscr{C}}$. Using this notation, one may take the liberty of writing

$$\begin{aligned} H_E^s(x_0) &= \frac{1}{2} \int_{\mathbb{R}^n} \frac{\tilde{\chi}_E(x_0 + y) + \tilde{\chi}_E(x_0 - y)}{|y|^{n+2s}} \, dy \\ &= \frac{1}{2} \int_{\mathbb{R}^n} \frac{\tilde{\chi}_E(x_0 + y) + \tilde{\chi}_E(x_0 - y) - 2\tilde{\chi}_E(x_0)}{|y|^{n+2s}} \, dy \\ &= \frac{-(-\Delta)^s \tilde{\chi}_E(x_0)}{C(n, s)}, \end{aligned}$$

using the notation of (1.1). Using this suggestive representation, the Euler-Lagrange equation in (5.5) becomes

$$(-\Delta)^s \tilde{\chi}_E = 0 \text{ along } \partial E.$$

We refer to [3] for further details on this argument.

It is also worth recalling that the nonlocal perimeter functionals find applications in motions of fronts by nonlocal mean curvature (see e.g. [32, 40, 91]), problems in which aggregating and disaggregating terms compete towards an equilibrium (see e.g. [78] and [56]) and nonlocal free boundary problems (see e.g. [29] and [61]). See also [106] and [139] for results related to this type of problems.

In the classical case of the local perimeter functional, it is known that minimal surfaces are smooth in dimension $n \leq 7$. Moreover, if $n \geq 8$ minimal surfaces are smooth except on a small singular set of Hausdorff dimension $n - 8$. Furthermore, minimal surfaces that are graphs are called minimal graphs, and they reduce to hyperplanes if $n \leq 8$ (this is called the Bernstein property, which was also discussed at the beginning of the Chap. 4). If $n \geq 9$, there exist global minimal graphs that are not affine (see e.g. [86]).

Differently from the classical case, the regularity theory for *s*-minimizers is still quite open. We present here some of the partial results obtained in this direction:

Theorem 5.3 *In the plane, s-minimal sets are smooth. More precisely:*

(a) *If E is an s-minimal set in $\Omega \subset \mathbb{R}^2$, then $\partial E \cap \Omega$ is a C^∞-curve.*
(b) *Let E be s-minimal in $\Omega \subset \mathbb{R}^n$ and let $\Sigma_E \subset \partial E \cap \Omega$ denote its singular set. Then $\mathscr{H}^d(\Sigma_E) = 0$ for any $d > n - 3$.*

See [124] for the proof of this results (as a matter of fact, in [124] only $C^{1,\alpha}$ regularity is proved, but then [9] proved that *s*-minimal sets with $C^{1,\alpha}$-boundary are automatically C^∞). Further regularity results of the *s*-minimal surfaces can be found in [35]. There, a regularity theory when *s* is near $\dfrac{1}{2}$ is stated, as we see in the following Theorem:

Theorem 5.4 *There exists $\epsilon_0 \in \left(0, \dfrac{1}{2}\right)$ such that if $s \geq \dfrac{1}{2} - \epsilon_0$, then*

(a) *if $n \leq 7$, any s-minimal set is of class C^∞,*
(b) *if $n = 8$ any s-minimal surface is of class C^∞ except, at most, at countably many isolated points,*
(c) *any s-minimal surface is of class C^∞ outside a closed set Σ of Hausdorff dimension $n - 8$.*

5.1 Graphs and *s*-Minimal Surfaces

We will focus the upcoming material on two interesting results related to graphs: a Bernstein type result, namely the property that an *s*-minimal graph in $\mathbb{R}^{n+1}$ is flat (if no singular cones exist in dimension n); we will then prove that an *s*-minimal surface whose prescribed data is a subgraph, is itself a subgraph.

The first result is the following theorem:

Theorem 5.1.1 *Let $E = \{(x, t) \in \mathbb{R}^n \times \mathbb{R} \text{ s.t. } t < u(x)\}$ be an s-minimal graph, and assume there are no singular cones in dimension n (that is, if $\mathscr{K} \subset \mathbb{R}^n$ is an s-minimal cone, then $\mathscr{K}$ is a half-space). Then u is an affine function (thus E is a half-space).*

To be able to prove Theorem 5.1.1, we recall some useful auxiliary results. In the following lemma we state a dimensional reduction result (see Theorem 10.1 in [28]).

Lemma 5.1 *Let $E = F \times \mathbb{R}$. Then if E is s-minimal if and only if F is s-minimal.*

We define then the blow-up and blow-down of the set E are, respectively

$$E_0 := \lim_{r \to 0} E_r \quad \text{and} \quad E_\infty := \lim_{r \to +\infty} E_r, \quad \text{where} \quad E_r = \frac{E}{r}.$$

A first property of the blow-up of E is the following (see Lemma 3.1 in [79]).

Lemma 5.2 *If E_∞ is affine, then so is E.*

We recall also a regularity result for the s-minimal surfaces (see [79] and [9] for details and proof).

Lemma 5.3 *Let E be s-minimal. Then:*

(a) If E is Lipschitz, then E is $C^{1,\alpha}$.
(b) If E is $C^{1,\alpha}$, then E is C^∞.

We give here a sketch of the proof of Theorem 5.1.1 (see [79] for all the details).

Proof (Sketch of the proof of Theorem 5.1.1) If $E \subset \mathbb{R}^{n+1}$ is an s-minimal graph, then the blow-down E_∞ is an s-minimal cone (see Theorem 9.2 in [28] for the proof of this statement). By applying the dimensional reduction argument in Lemma 5.1 we obtain an s-minimal cone in dimension n. According to the assumption that no singular s-minimal cones exist in dimension n, it follows that necessarily E_∞ can be singular only at the origin.

We consider a bump function $w_0 \in C^\infty(\mathbb{R}, [0, 1])$ such that

$$w_0(t) = 0 \text{ in } \left(-\infty, \frac{1}{4} \right) \cup \left(\frac{3}{4}, +\infty \right)$$

$$w_0(t) = 1 \text{ in } \left(\frac{2}{5}, \frac{3}{5} \right)$$

$$w(t) = w_0(|t|).$$

The blow-down of E is

$$E_\infty = \{(x', x_{n+1}) \text{ s.t. } x_{n+1} \le u_\infty(x')\}.$$

For a fixed $\sigma \in \partial B_1$, let

$$F_t := \{(x', x_{n+1}) \text{ s.t. } x_{n+1} \le u_\infty(x' + t\theta w(x')\sigma) - t\}$$

be a family of sets, where $t \in (0, 1)$ and $\theta > 0$. Then for θ small, we have that

$$F_1 \text{ is below } E_\infty. \qquad (5.7)$$

Indeed, suppose by contradiction that this is not true. Then, there exists $\theta_k \to 0$ such that

$$u_\infty\big(x_k' + \theta_k w(x_k')\sigma\big) - 1 \geq u_\infty(x_k'). \qquad (5.8)$$

But $x_k' \in \operatorname{supp} w$, which is compact, therefore $x_\infty' := \lim_{k \to +\infty} x_k'$ belongs to the support of w, and $w(x_\infty')$ is defined. Then, by sending $k \to +\infty$ in (5.8) we have that

$$u_\infty(x_\infty') - 1 \geq u_\infty(x_\infty'),$$

which is a contradiction. This establishes (5.7).

Now consider the smallest $t_0 \in (0, 1)$ for which F_t is below E_∞. Since E_∞ is a graph, then F_{t_0} touches E_∞ from below in one point $X_0 = (x_0', x_{n+1}^0)$, where $x_0' \in \operatorname{supp} w$. Since E_∞ is s-minimal, we have that the nonlocal mean curvature (defined in (5.6)) of the boundary is null. Also, since F_{t_0} is a C^2 diffeomorphism of E_∞ we have that

$$H^s_{F_{t_0}}(p) \simeq \theta t_0, \qquad (5.9)$$

and there is a region where E_∞ and F_{t_0} are well separated by t_0, thus

$$\big|(E_\infty \setminus F_{t_0}) \cap (B_3 \setminus B_2)\big| \geq ct_0,$$

for some $c > 0$. Therefore, we see that

$$H^s_{F_{t_0}}(p) = H^s_{F_{t_0}}(p) - H^s_E(p) \geq ct_0.$$

This and (5.9) give that $\theta t_0 \geq ct_0$, for some $c > 0$ (up to renaming it). If θ is small enough, this implies that $t_0 = 0$.

In particular, we have proved that there exists $\theta > 0$ small enough such that, for any $t \in (0, 1)$ and any $\sigma \in \partial B_1$, we have that

$$u_\infty\big(x' + t\theta w(x')\sigma\big) - t \leq u_\infty(x').$$

This implies that

$$\frac{u_\infty\big(x' + t\theta w(x')\sigma\big) - u_\infty(x')}{t\theta} \leq \frac{1}{\theta},$$

hence, letting $t \to 0$, we have that

$$\nabla u_\infty(x')w(x')\sigma \le \frac{1}{\theta}, \text{ for any } x \in \mathbb{R}^n \setminus \{0\}, \text{ and } \sigma \in B_1.$$

We recall now that $w = 1$ in $B_{3/5} \setminus B_{2/5}$ and σ is arbitrary in ∂B_1. Hence, it follows that

$$|\nabla u_\infty(x)| \le \frac{1}{\theta}, \text{ for any } x \in B_{3/5} \setminus B_{2/5}.$$

Therefore u_∞ is globally Lipschitz. By the regularity statement in Lemma 5.3, we have that u_∞ is C^∞. This says that u is smooth also at the origin, hence (being a cone) it follows that E_∞ is necessarily a half-space. Then by Lemma 5.2, we conclude that E is a half-space as well.

We introduce in the following theorem another interesting property related to s-minimal surfaces, in the case in which the fixed given data outside a domain is a subgraph. In that case, the s-minimal surface itself is a subgraph. Indeed:

Theorem 5.1.2 *Let Ω_0 be an open and bounded subset of $\mathbb{R}^{n-1}$ with boundary of class $C^{1,1}$ and let $\Omega := \Omega_0 \times \mathbb{R}$. Let E be an s-minimal set in Ω. Assume that*

$$E \setminus \Omega = \{x_n < u(x'), x' \in \mathbb{R}^{n-1} \setminus \Omega_0\} \qquad (5.10)$$

for some continuous function $u: \mathbb{R}^{n-1} \to \mathbb{R}$. Then

$$E \cap \Omega = \{x_n < v(x'), x' \in \Omega_0\}$$

for some function $v: \mathbb{R}^{n-1} \to \mathbb{R}$.

The reader can see [64], where this theorem and the related results are proved; here, we only state the preliminary results needed for our purposes and focus on the proof of Theorem 5.1.2. The proof relies on a sliding method, more precisely, we take a translation of E in the nth direction, and move it until it touches E from above. If the set $E \cap \Omega$ is a subgraph, then, up to a set of measure 0, the contact between the translated E and E, will be E itself.

However, since we have no information on the regularity of the minimal surface, we need at first to "regularize" the set by introducing the notions of supconvolution and subconvolution. With the aid of a useful result related to the sub/supconvolution of an s-minimal surface, we proceed then with the proof of the Theorem 5.1.2.

The supconvolution of a set $E \subseteq \mathbb{R}^n$ (Fig. 5.2) is given by

$$E_\delta^\# := \bigcup_{x \in E} \overline{B_\delta(x)}.$$

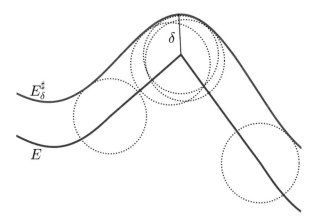

Fig. 5.2 The supconvolution of a set

In an equivalent way, the supconvolution can be written as

$$E_\delta^\sharp = \bigcup_{\substack{v \in \mathbb{R}^n \\ |v| \le \delta}} (E + v).$$

Indeed, we consider $\delta > 0$ and an arbitrary $x \in E$. Let $y \in \overline{B_\delta(x)}$ and we define $v := y - x$. Then

$$|v| \le |y - x| \le \delta \quad \text{and} \quad y = x + v \in E + v.$$

Therefore $\overline{B_\delta(x)} \subseteq E + v$ for $|v| \le \delta$. In order to prove the inclusion in the opposite direction, one notices that taking $y \in E + v$ with $|v| \le \delta$ and defining $x := y - v$, it follows that

$$|x - y| = |v| \le \delta.$$

Moreover, $x \in (E + v) - v = E$ and the inclusion $E + v \in \overline{B_\delta(x)}$ is proved.

On the other hand, the subconvolution is defined as

$$E_\delta^\flat := \mathbb{R}^n \setminus \left((\mathbb{R}^n \setminus E)_\delta^\sharp \right).$$

Now, the idea is that the supconvolution of E is a regularized version of E whose nonlocal minimal curvature is smaller than the one of E, i.e.:

$$\int_{\mathbb{R}^n} \frac{\chi_{\mathbb{R}^n \setminus E_\delta^\sharp}(y) - \chi_{E_\delta^\sharp}(y)}{|x - y|^{n+2s}} \, dy \le \int_{\mathbb{R}^n} \frac{\chi_{\mathbb{R}^n \setminus E}(y) - \chi_E(y)}{|\tilde{x} - y|^{n+2s}} \, dy \le 0, \tag{5.11}$$

for any $x \in \partial E_\delta^\sharp$, where $\tilde{x} := x - v \in \partial E$ for some $v \in \mathbb{R}^n$ with $|v| = \delta$. Then, by construction, the set $E + v$ lies in $E_\delta^\sharp$, and this implies (5.11).

Similarly, one has that the opposite inequality holds for the subconvolution of E, for any $x \in \partial E_\delta^\flat$

$$\int_{\mathbb{R}^n} \frac{\chi_{\mathbb{R}^n \setminus E_\delta^\flat}(y) - \chi_{E_\delta^\flat}(y)}{|x - y|^{n+2s}} \, dy \geq 0, \tag{5.12}$$

By (5.11) and (5.12), we obtain:

Proposition 5.1.3 *Let E be an s-minimal set in Ω. Let $p \in \partial E_\delta^\sharp$ and assume that* $\overline{B_\delta(p)} \subseteq \Omega$. *Assume also that $E_\delta^\sharp$ is touched from above by a translation of $E_\delta^\flat$, namely there exists $\omega \in \mathbb{R}^n$ such that*

$$E_\delta^\sharp \subseteq E_\delta^\flat + \omega$$

and

$$p \in (\partial E_\delta^\sharp) \cap (\partial E_\delta^\flat + \omega).$$

Then

$$E_\delta^\sharp = E_\delta^\flat + \omega.$$

Proof (Proof of Theorem 5.1.2) One first remark is that the s-minimal set does not have spikes which go to infinity: more precisely, one shows that

$$\Omega_0 \times (-\infty, -M) \subseteq E \cap \Omega \subseteq \Omega_0 \times (-\infty, M) \tag{5.13}$$

for some $M \geq 0$. The proof of (5.13) can be performed by sliding horizontally a large ball, see [64] for details.

After proving (5.13), one can deal with the core of the proof of Theorem 5.1.2. The idea is to slide E from above until it touches itself and analyze what happens at the contact points. For simplicity, we will assume here that the function u is uniformly continuous (if u is only continuous, the proof needs to be slightly modified since the subconvolution and supconvolution that we will perform may create new touching points at infinity). At this purpose, we consider $E_t = E + te_n$ for $t \geq 0$. Notice that, by (5.13), if $t \geq 2M$, then $E \subseteq E_t$. Let then t be the smallest for which the inclusion $E \subseteq E_t$ holds. We claim that $t = 0$. If this happens, one may consider

$$v = \inf\{\tau \text{ s.t. } (x, \tau) \in E^\mathscr{C}\}$$

and, up to sets of measure 0, $E \cap \Omega_0$ is the subgraph of v.

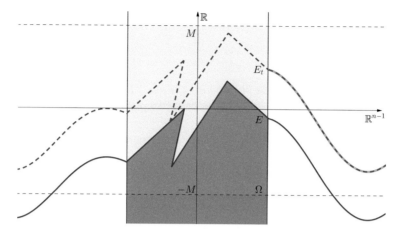

Fig. 5.3 Sliding E until it touches itself at an interior point

The proof is by contradiction, so let us assume that $t > 0$. According to (5.10), the set $E \setminus \Omega$ is a subgraph, hence the contact points between ∂E and ∂E_t must lie in $\overline{\Omega}_0 \times \mathbb{R}$. Namely, only two possibilities may occur: the contact point is interior (it belongs to $\Omega_0 \times \mathbb{R}$), or it is at the boundary (on $\partial\Omega_0 \times \mathbb{R}$). So, calling p the contact point, one may have[1] that

$$\text{either } p \in \Omega_0 \times \mathbb{R} \quad \text{or} \qquad\qquad (5.14)$$

$$p \in \partial\Omega_0 \times \mathbb{R}. \qquad\qquad (5.15)$$

We deal with the first case in (5.14) (an example of this behavior is depicted in Fig. 5.3). We consider $E_\delta^\sharp$ and $E_\delta^\flat$ to be the supconvolution, respectively the subconvolution of E. We then slide the subconvolution until it touches the supconvolution. More precisely, let $\tau > 0$ and we take a translation of the subconvolution, $E_\delta^\flat + \tau e_n$. For τ large, we have that $E_\delta^\sharp \subseteq E_\delta^\flat + \tau e_n$ and we consider τ_δ to be the smallest for

---

[1] As a matter of fact, the number of contact points may be higher than one, and even infinitely many contact points may arise. So, to be rigorous, one should distinguish the case in which all the contact points are interior and the case in which at least one contact point lies on the boundary.

Moreover, since the surface may have vertical portions along the boundary of the domain, one needs to carefully define the notion of contact points (roughly speaking, one needs to take a definition for which the vertical portions which do not prevent the sliding are not in the contact set).

Finally, in case the contact points are all interior, it is also useful to perform the sliding method in a slightly reduced domain, in order to avoid that the supconvolution method produces new contact points at the boundary (which may arise from vertical portions of the surfaces).

Since we do not aim to give a complete proof of Theorem 5.1.2 here, but just to give the main ideas and underline the additional difficulty, we refer to [64] for the full details of these arguments.

which such inclusion holds. We have (since t is positive by assumption) that

$$\tau_\delta \geq \frac{t}{2} > 0.$$

Moreover, for δ small, the sets $\partial E_\delta^\sharp$ and $\partial (E_\delta^\flat + \tau_\delta e_n)$ have a contact point which, according to (5.14), lies in $\Omega_0 \times \mathbb{R}$. Let p_δ be such a point, so we may write

$$p_\delta \in (\partial E_\delta^\sharp) \cap \partial (E_\delta^\flat + \tau_\delta e_n) \quad \text{and} \quad p_\delta \in \Omega_0 \times \mathbb{R}.$$

Then, for δ small (notice that $\overline{B_\delta(p)} \subseteq \Omega$), Proposition 5.1.3 yields that

$$E_\delta^\sharp = E_\delta^\flat + \tau_\delta e_n.$$

Considering δ arbitrarily small, one obtains that

$$E = E + \tau_0 e_n, \quad \text{with} \quad \tau_0 > 0.$$

But E is a subgraph outside of Ω, and this provides a contradiction. Hence, the claim that $t = 0$ is proved.

Let us see that we also obtain a contradiction when supposing that $t > 0$ and that the second case (5.15) holds. Let

$$p = (p', p_n) \quad \text{and} \quad p \in (\partial E) \cap (\partial E_t).$$

Now, if one takes sequences $a_k \in \partial E$ and $b_k \in \partial E_t$, both that tend to p as k goes to infinity, since $E \setminus \Omega$ is a subgraph and $t > 0$, necessarily a_k, b_k belong to Ω. Hence

$$p \in \overline{(\partial E) \cap \Omega} \cap \overline{(\partial E_t) \cap \Omega}. \tag{5.16}$$

Thanks to Definition 2.3 in [28], one obtains that E is a variational subsolution in a neighborhood of p. In other words, if $A \subseteq E \cap \Omega$ and $p \in \overline{A}$, then

$$0 \geq \operatorname{Per}_s(E, \Omega) - \operatorname{Per}_s(E \setminus A, \Omega) = I(A, E^{\mathscr{C}}) - I(A, E \setminus A)$$

(we recall the definition of I in (5.1) and of the fractional perimeter Per_s in (5.2)). According to Theorem 5.1 in [28], this implies in a viscosity sense (i.e. if E is touched at p from outside by a ball), that

$$\int_{\mathbb{R}^n} \frac{\chi_E(y) - \chi_{\mathbb{R}^n \setminus E}(y)}{|p - y|^{n+2s}} \, dy \geq 0. \tag{5.17}$$

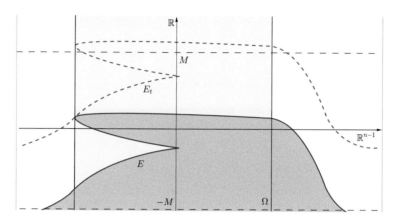

Fig. 5.4 Sliding E until it touches itself at a boundary point

In order to obtain an estimate on the fractional mean curvature in the strong sense, we consider the translation of the point p as follows:

$$p_t = p - te_n = (p', p_n - t) = (p', p_{n,t}).$$

Since $t > 0$, one may have that either $p_n \neq u(p')$, or $p_{n,t} \neq u(p')$.

These two possibilities can be dealt with in a similar way, so we just continue with the proof in the case $p_n \neq u(p')$ (as is also exemplified in Fig. 5.4). Taking $r > 0$ small, the set $B_r(p) \setminus \Omega$ is contained entirely in E or in its complement. Moreover, one has from [27] that $\partial E \cap B_r(p)$ is a $C^{1, \frac{1}{2}+s}$-graph in the direction of the normal to Ω at p. That is: in Fig. 5.4 the set E is $C^{1, \frac{1}{2}+s}$, hence in the vicinity of $p = (p', p_n)$, it appears to be sufficiently smooth.

So, let $\nu(p) = (\nu'(p), \nu_n(p))$ be the normal in the interior direction, then up to a rotation and since Ω is a cylinder (hence $\nu_n(p) = 0$), we can write $\nu(p) = e_1$. Therefore, there exists a function Ψ of class $C^{1, \frac{1}{2}+s}$ such that $p_1 = \Psi(p_2, \ldots, p_n)$ and, in the vicinity of p, we can write ∂E as the graph $G = \{x_1 = \Psi(x_2, \ldots, x_n)\}$.

Given (5.16), we deduce that there exists a sequence $p_k \in G$ such that $p_k \in \Omega$ and $p_k \to p$ as $k \to \infty$. From this it follows that there exists a sequence of points $p_k \to p$ such that

$$\partial E \text{ in the vicinity of } p_k \text{ is a graph of class } C^{1, \frac{1}{2}+s} \tag{5.18}$$

and

$$\int_{\mathbb{R}^n} \frac{\chi_E(y) - \chi_{E^{\mathscr{C}}}(y)}{|p_k - y|^{n+2s}} \, dy = 0. \tag{5.19}$$

From (5.18) and (5.19), and using a pointwise version of the Euler-Lagrange equation (see [64] for details), we have that

$$\int_{\mathbb{R}^n} \frac{\chi_E(y) - \chi_{E^\mathscr{C}}(y)}{|p - y|^{n+2s}} \, dy = 0.$$

Now, $E \subset E_t$ for t strictly positive, hence

$$\int_{\mathbb{R}^n} \frac{\chi_{E_t}(y) - \chi_{E^\mathscr{C}}(y)}{|p - y|^{n+2s}} \, dy > 0. \tag{5.20}$$

Moreover, we have that the set $\partial E_t \cap B_{\frac{r}{4}}(p)$ must remain on one side of the graph G, namely one could have that

$$E_t \cap B_{\frac{r}{4}}(p) \subseteq \{x_1 \leq \Psi(x_2, \ldots, x_n)\} \text{ or}$$

$$E_t \cap B_{\frac{r}{4}}(p) \supseteq \{x_1 \geq \Psi(x_2, \ldots, x_n)\}.$$

Given again (5.16), we deduce that there exists a sequence $\tilde{p}_k \in \partial E_t \cap \Omega$ such that $\tilde{p}_k \to p$ as $k \to \infty$ and $\partial E_t \cap \Omega$ in the vicinity of $\tilde{p}_k$ is touched by a surface lying in E_t, of class $C^{1, \frac{1}{2} + s}$. Then

$$\int_{\mathbb{R}^n} \frac{\chi_{E_t}(y) - \chi_{E_t^\mathscr{C}}(y)}{|\tilde{p}_k - y|^{n+2s}} \, dy \leq 0.$$

Hence, making use of a pointwise version of the Euler-Lagrange equation (see [64] for details), we obtain that

$$\int_{\mathbb{R}^n} \frac{\chi_{E_t}(y) - \chi_{E_t^\mathscr{C}}(y)}{|p - y|^{n+2s}} \, dy \leq 0.$$

But this is a contradiction with (5.20), and this concludes the proof of Theorem 5.1.2.

On the one hand, one may think that Theorem 5.1.2 has to be well-expected. On the other hand, it is far from being obvious, not only because the proof is not trivial, but also because the statement itself almost risks to be false, especially at the boundary. Indeed we will see in Theorem 5.3.2 that the graph property is close to fail at the boundary of the domain, where the s-minimal surfaces may present vertical tangencies and stickiness phenomena (see Fig. 5.11).

5.2 Non-existence of Singular Cones in Dimension 2

We now prove the non-existence of singular s-minimal cones in dimension 2, as stated in the next result (from this, the more general statement in Theorem 5.3 follows after a blow-up procedure):

Theorem 5.2.1 *If E is an s-minimal cone in $\mathbb{R}^2$, then E is a half-plane.*

We remark that, as a combination of Theorems 5.1.1 and 5.2.1, we obtain the following result of Bernstein type:

Corollary 5.2.2 *Let $E = \{(x, t) \in \mathbb{R}^n \times \mathbb{R} \text{ s.t. } t < u(x)\}$ be an s-minimal graph, and assume that $n \in \{1, 2\}$. Then u is an affine function.*

Let us first consider a simple example, given by the cone in the plane

$$\mathscr{K} := \left\{(x, y) \in \mathbb{R}^2 \text{ s.t. } y^2 > x^2\right\},$$

see Fig. 5.5.

Proposition 5.2.3 *The cone $\mathscr{K}$ depicted in Fig. 5.5 is not s-minimal in $\mathbb{R}^2$.*

Notice that, by symmetry, one can prove that $\mathscr{K}$ satisfies (5.5) (possibly in the viscosity sense). On the other hand, Proposition 5.2.3 gives that $\mathscr{K}$ is not s-minimal. This, in particular, provides an example of a set that satisfies the Euler-Lagrange equation in (5.5), but is not s-minimal (i.e., the Euler-Lagrange equation in (5.5) is implied by, but not necessarily equivalent to, the s-minimality property).

Proof (Proof of Proposition 5.2.3) The proof of the non-minimality of $\mathscr{K}$ is due to an original idea by Luis Caffarelli.

Suppose by contradiction that the cone $\mathscr{K}$ is minimal in $\mathbb{R}^2$. We add to $\mathscr{K}$ a small square adjacent to the origin (see Fig. 5.6), and call $\mathscr{K}'$ the set obtained. Then $\mathscr{K}$ and $\mathscr{K}'$ have the same s-perimeter. This is due to the interactions considered in

Fig. 5.5 The cone $\mathscr{K}$

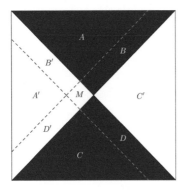

Fig. 5.6 Interaction of M with $A, B, C, D, A', B', C', D'$

the s-perimeter functional and the unboundedness of the regions. We remark that in Fig. 5.6 we represent bounded regions, of course, sets A, B, C, D, A', B', C' and D' are actually unbounded.

Indeed, we notice that in the first image, the white square M interacts with the dark regions A, B, C, D, while in the second the now dark square M interacts with the regions A', B', C', D', and all the other interactions are unmodified. Therefore, the difference between the s-perimeter of $\mathscr{K}$ and that of $\mathscr{K}'$ consists only of the interactions $I(A, M)+I(B, M)+I(C, M)+I(D, M)-I(A', M)-I(B', M)-I(C', M)-I(D', M)$. But $A \cup B = A' \cup B'$ and $C \cup D = C' \cup D'$ (since these sets are all unbounded), therefore the difference is null, and the s-perimeter of $\mathscr{K}$ is equal to that of $\mathscr{K}'$. Consequently, $\mathscr{K}'$ is also s-minimal, and therefore it satisfies the Euler-Lagrange equation in (5.5) at the origin. But this leads to a contradiction, since the dark region now contributes more than the white one, namely

$$\int_{\mathbb{R}^2} \frac{\chi_{\mathscr{K}'}(y) - \chi_{\mathbb{R}^2 \setminus \mathscr{K}'}(y)}{|y|^{2+s}}\, dy > 0.$$

Thus $\mathscr{K}$ cannot be s-minimal, and this concludes our proof.

This geometric argument cannot be extended to a more general case (even, for instance, to a cone in $\mathbb{R}^2$ made of many sectors, see Fig. 5.7). As a matter of fact, the proof of Theorem 5.2.1 will be completely different than the one of Proposition 5.2.3 and it will rely on an appropriate domain perturbation argument.

The proof of Theorem 5.2.1 that we present here is actually different than the original one in [124]. Indeed, in [124], the result was proved by using the harmonic extension for the fractional Laplacian. Here, the extension will not be used; furthermore, the proof follows the steps of Theorem 4.2.1 and we will recall here just the main ingredients.

Fig. 5.7 Cone in $\mathbb{R}^2$

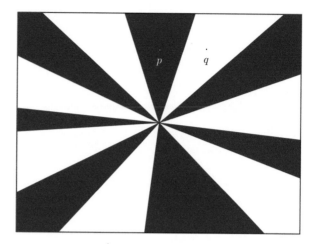

Proof (Proof of Theorem 5.2.1)

The idea of the proof is the following: if $E \subset \mathbb{R}^2$ is an s-minimal cone, then let $\tilde{E}$ be a perturbation of the set E which coincides with a translation of E in $B_{R/2}$ and with E itself outside B_R. Then the difference between the energies of $\tilde{E}$ and E tends to 0 as $R \to +\infty$. This implies that also the energy of $E \cap \tilde{E}$ is arbitrarily close to the energy of E. On the other hand if E is not a half-plane, the set $\tilde{E} \cap E$ can be modified locally to decrease its energy by a fixed small amount and we reach a contradiction.

The details of the proof go as follows. Let

$$u := \chi_E - \chi_{\mathbb{R}^2 \setminus E}.$$

From definition (4.35) we have that

$$u(B_R, B_R) = 2I(E \cap B_R, B_R \setminus E)$$

and

$$u(B_R, B_R^{\mathscr{C}}) = I(B_R \cap E, E^{\mathscr{C}} \setminus B_R) + I(B_R \setminus E, E \setminus B_R),$$

thus

$$\mathrm{Per}_s(E, B_R) = \mathscr{K}_R(u), \tag{5.21}$$

where $\mathscr{K}_R(u)$ is given in (4.33) and $\mathrm{Per}_s(E, B_R)$ is the s-perimeter functional defined in (5.2). Then E is s-minimal if u is a minimizer of the energy $\mathscr{K}_R$ in any ball B_R, with $R > 0$.

Now, we argue by contradiction, and suppose that E is an s-minimal cone different from the half-space. Up to rotations, we may suppose that a sector of E

has an angle smaller than π and is bisected by e_2. Thus there exists $M \geq 1$ and $p \in E \cap B_M$ on the e_2-axis such that $p \pm e_1 \in \mathbb{R}^2 \setminus E$ (see Fig. 5.7).

We take $\varphi \in C_0^\infty(B_1)$, such that $\varphi(x) = 1$ in $B_{1/2}$. For R large (say $R > 8M$), we define

$$\Psi_{R,+}(y) := y + \varphi\left(\frac{y}{R}\right)e_1.$$

We point out that, for R large, $\Psi_{R,+}$ is a diffeomorphism on $\mathbb{R}^2$.

Furthermore, we define $u_R^+(x) := u(\Psi_{R,+}^{-1}(x))$. Then

$$u_R^+(y) = u(y - e_1) \quad \text{for } p \in B_{2M}$$
$$\text{and} \quad u_R^+(y) = u(y) \quad \text{for } p \in \mathbb{R}^2 \setminus B_R.$$

We recall the estimate obtained in (4.37), that, combined with the minimality of u, gives

$$\mathcal{H}_R(u_R^+) - \mathcal{H}_R(u) \leq \frac{C}{R^2}\mathcal{H}_R(u).$$

But u is a minimizer in any ball, and by the energy estimate in Theorem 4.1.2 we have that

$$\mathcal{H}_R(u_R^+) - \mathcal{H}_R(u) \leq CR^{-2s}.$$

This implies that

$$\lim_{R\to+\infty} \mathcal{H}_R(u_R^+) - \mathcal{H}_R(u) = 0. \tag{5.22}$$

Let now

$$v_R(x) := \min\{u(x), u_R^+(x)\} \quad \text{and} \quad w_R(x) := \max\{u(x), u_R^+(x)\}.$$

We claim that v_R is not identically u nor u_R^+. Indeed

$$u_R^+(p) = u(p - e_1) = (\chi_E - \chi_{\mathbb{R}^2\setminus E})(p - e_1) = -1 \quad \text{and}$$
$$u(p) = (\chi_E - \chi_{\mathbb{R}^2\setminus E})(p) = 1.$$

On the other hand,

$$u_R^+(p + e_1) = u(p) = 1 \quad \text{and}$$
$$u(p + e_1) = (\chi_E - \chi_{\mathbb{R}^2\setminus E})(p + e_1) = -1.$$

By the continuity of u and u_R^+, we obtain that

$$v_R = u_R^+ < u \text{ in a neighborhood of } p \tag{5.23}$$

and

$$v_R = u < u_R^+ \text{ in a neighborhood of } p + e_1. \tag{5.24}$$

Now, by the minimality property of u,

$$\mathcal{K}_R(u) \leq \mathcal{K}_R(v_R).$$

Moreover (see e.g. formula (38) in [114]),

$$\mathcal{K}_R(v_R) + \mathcal{K}_R(w_R) \leq \mathcal{K}_R(u) + \mathcal{K}_R(u_R^+).$$

The latter two formulas give that

$$\mathcal{K}_R(v_R) \leq \mathcal{K}_R(u_R^+). \tag{5.25}$$

We claim that

$$v_R \text{ is not minimal for } \mathcal{K}_{2M} \tag{5.26}$$

with respect to compact perturbations in B_{2M}. Indeed, assume by contradiction that v_R is minimal, then in B_{2M} both v_R and u would satisfy the same equation. Recalling (5.24) and applying the Strong Maximum Principle, it follows that $u = v_R$ in B_{2M}, which contradicts (5.23). This establishes (5.26).

Now, we consider a minimizer u_R^* of $\mathcal{K}_{2M}$ among the competitors that agree with v_R outside B_{2M}. Therefore, we can define

$$\delta_R := \mathcal{K}_{2M}(v_R) - \mathcal{K}_{2M}(u_R^*).$$

In light of (5.26), we have that $\delta_R > 0$.

The reader can now compare Step 3 in the proof of Theorem 4.2.1. There we proved that

$$\delta_R \text{ remains bounded away from zero as } R \to +\infty. \tag{5.27}$$

Furthermore, since u_R^* and v_R agree outside B_{2M} we obtain that

$$\mathcal{K}_R(u_R^*) + \delta_R = \mathcal{K}_R(v_R).$$

Using this, (5.25) and the minimality of u, we obtain that

$$\delta_R = \mathcal{K}_R(v_R) - \mathcal{K}_R(u_R^*) \le \mathcal{K}_R(u_R^+) - \mathcal{K}_R(u).$$

Now we send R to infinity, recall (5.22) and (5.27), and we reach a contradiction. Thus, E is a half-space, and this concludes the proof of Theorem 5.2.1.

As already mentioned, the regularity theory for s-minimal sets is still widely open. Little is known beyond Theorems 5.3 and 5.4, so it would be very interesting to further investigate the regularity of s-minimal surfaces in higher dimension and for small s.

It is also interesting to recall that if the s-minimal surface E is a subgraph of some function $u : \mathbb{R}^{n-1} \to \mathbb{R}$ (at least in the vicinity of some point $x_0 = (x_0', u(x_0')) \in \partial E$) then the Euler-Lagrange (5.5) can be written directly in terms of u. For instance (see formulas (49) and (50) in [9]), under appropriate smoothness assumptions on u, formula (5.5) reduces to

$$0 = \int_{\mathbb{R}^n} \frac{\chi_{\mathbb{R}^n \setminus E}(x_0 + y) - \chi_E(x_0 + y)}{|y|^{n+2s}} \, dy$$

$$= \int_{\mathbb{R}^{n-1}} F\left(\frac{u(x_0' + y') - u(x_0')}{|y'|}\right) \frac{\zeta(y')}{|y'|^{n-1+2s}} \, dy' + \Psi(x_0'),$$

for suitable F and Ψ, and a cut-off function ζ supported in a neighborhood of x_0'.

Regarding the regularity problems of the s-minimal surfaces, let us mention the recent papers [47] and [48]. Among other very interesting results, it is proved there that suitable singular cones of symmetric type are unstable up to dimension 6 but become stable in dimension 7 for small s (these cones can be seen as the nonlocal analogue of the Lawson cones in the classical minimal surface theory, and the stability property is in principle weaker than minimality, since it deals with the positivity of the second order derivative of the functional).

This phenomenon may suggest the conjecture that the s-minimal surfaces may develop singularities in dimension 7 and higher when s is sufficiently small.

In [48], interesting examples of surfaces with vanishing nonlocal mean curvature are provided for s sufficiently close to $1/2$. Remarkably, the surfaces in [48] are the nonlocal analogues of the catenoids, but, differently from the classical case (in which catenoids grow logarithmically), they approach a singular cone at infinity, see Fig. 5.8.

Also, these nonlocal catenoids are highly unstable from the variational point of view, since they possess infinite Morse index (differently from the standard catenoid, which has Morse index equal to one, i.e. it is, roughly speaking, a minimizer in any functional direction with the exception of one).

Moreover, in [48], there are also examples of surfaces with vanishing nonlocal mean curvature that can be seen as the nonlocal analogues of two parallel hyperplanes. Namely, for s sufficiently close to $1/2$, there exists a surface of revolution made of two sheets which are the graph of a radial function $f = \pm f(r)$. When r

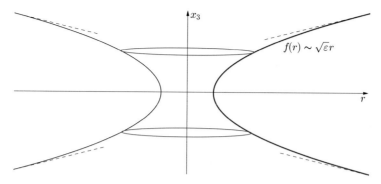

Fig. 5.8 A nonlocal catenoid

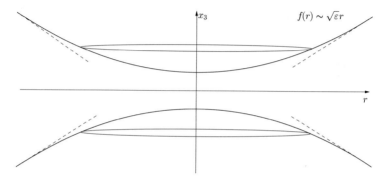

Fig. 5.9 A two-sheet surface with vanishing fractional mean curvature

is small, f is of the order of $1 + (\frac{1}{2} - s)r^2$, but for large r it becomes of the order of $\sqrt{\frac{1}{2} - s} \cdot r$. That is, the two sheets "repel each other" and produce a linear growth at infinity. When s approaches $1/2$ the two sheets are locally closer and closer to two parallel hyperplanes, see Fig. 5.9.

The construction above may be extended to build families of surfaces with vanishing nonlocal mean curvature that can be seen as the nonlocal analogue of k parallel hyperplanes, for any $k \in \mathbb{N}$. These k-sheet surfaces can be seen as the bifurcation, as s is close to $1/2$, of the parallel hyperplanes $\{x_n = a_i\}$, for $i \in \{1, \dots, k\}$, where the parameters a_i satisfy the constraints

$$a_1 > \cdots > a_k, \qquad \sum_{i=1}^{k} a_i = 0 \qquad (5.28)$$

and the balancing relation

$$a_i = 2 \sum_{\substack{1 \leq j \leq n \\ j \neq i}} \frac{(-1)^{i+j+1}}{a_i - a_j}. \tag{5.29}$$

It is actually quite interesting to observe that solutions of (5.29) correspond to (nondegenerate) critical points of the functional

$$E(a_1, \dots, a_k) := \frac{1}{2} \sum_{i=1}^{k} a_i^2 + \sum_{\substack{1 \leq j \leq n \\ j \neq i}} (-1)^{i+j} \log |a_i - a_j|$$

among all the k-ples $(a_1, \dots, a_k)$ that satisfy (5.28).

These bifurcation techniques rely on a careful expansion of the fractional perimeter functional with respect to normal perturbations. That is, if E is a (smooth) set with vanishing fractional mean curvature, and h is a smooth and compactly supported perturbation, one can define, for any $t \in \mathbb{R}$,

$$E_h(t) := \{x + th(x)\nu(x), \ x \in \partial E\},$$

where $\nu(x)$ is the exterior normal of E at x. Then, the second variation of the perimeter of $E_h(t)$ at $t = 0$ is (up to normalization constants)

$$\int_{\partial E} \frac{h(y) - h(x)}{|x - y|^{n+2s}} \, d\mathcal{H}^{n-1}(y) + h(x) \int_{\partial E} \frac{(\nu(x) - \nu(y)) \cdot \nu(x)}{|x - y|^{n+2s}} \, d\mathcal{H}^{n-1}(y)$$

$$= \int_{\partial E} \frac{h(y) - h(x)}{|x - y|^{n+2s}} \, d\mathcal{H}^{n-1}(y) + h(x) \int_{\partial E} \frac{1 - \nu(x) \cdot \nu(y)}{|x - y|^{n+2s}} \, d\mathcal{H}^{n-1}(y).$$

Notice that the latter integral is non-negative, since $\nu(x) \cdot \nu(y) \leq 1$. The quantity above, in dependence of the perturbation h, is called, in jargon, "Jacobi operator". It encodes an important geometric information, and indeed, as $s \to 1/2$, it approaches the classical operator

$$\Delta_{\partial E} h + |A_{\partial E}|^2 h,$$

where $\Delta_{\partial E}$ is the Laplace-Beltrami operator along the hypersurface ∂E and $|A_{\partial E}|^2$ is the sum of the squares of the principal curvatures.

Other interesting sets that possess constant nonlocal mean curvature with the structure of onduloids have been recently constructed in [49] and [24]. This type of sets are periodic in a given direction and their construction has perturbative nature (indeed, the sets are close to a slab in the plane).

It is interesting to remark that the planar objects constructed in [24] have no counterpart in the local framework, since hypersurfaces of constant classical mean curvature with an onduloidal structure only exist in $\mathbb{R}^n$ with $n \geq 3$: once again, this is a typical nonlocal effect, in which the nonlocal mean curvature at a point is influenced by the global shape of the set.

While unbounded sets with constant nonlocal mean curvature and interesting geometric features have been constructed in [24, 48], the case of smooth and bounded sets is always geometrically trivial. As a matter of fact, it has been recently proved independently in [24] and [43] that bounded sets with smooth boundary and constant mean curvature are necessarily balls (this is the analogue of a celebrated result by Alexandrov for surfaces of constant classical mean curvature).

5.3 Boundary Regularity

The boundary regularity of the nonlocal minimal surfaces is also a very interesting, and surprising, topic. Indeed, differently from the classical case, nonlocal minimal surfaces do not always attain boundary data in a continuous way (not even in low dimension). A possible boundary behavior is, on the contrary, a combination of stickiness to the boundary and smooth separation from the adjacent portions. Namely, the nonlocal minimal surfaces may have a portion that sticks at the boundary and that separates from it in a $C^{1,\frac{1}{2}+s}$-way. As an example, we can consider, for any $\delta > 0$, the spherical cap

$$K_\delta := \left(B_{1+\delta} \setminus B_1 \right) \cap \{x_n < 0\},$$

and obtain the following stickiness result:

Theorem 5.3.1 *There exists $\delta_0 > 0$, depending on n and s, such that for any $\delta \in (0, \delta_0]$, we have that the s-minimal set in B_1 that coincides with K_δ outside B_1 is K_δ itself.*

That is, the s-minimal set with datum K_δ outside B_1 is empty inside B_1.

The stickiness property of Theorem 5.3.1 is depicted in Fig. 5.10.

Fig. 5.10 Stickiness
properties of Theorem 5.3.1

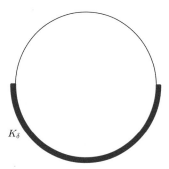

K_δ

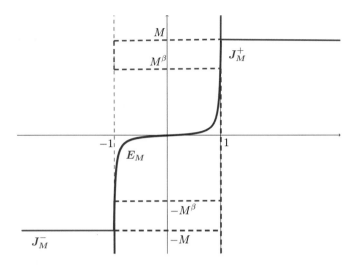

Fig. 5.11 Stickiness properties of Theorem 5.3.2

Other stickiness examples occur at the sides of slabs in the plane. For instance, given $M > 1$, one can consider the s-minimal set E_M in $(-1, 1) \times \mathbb{R}$ with datum outside $(-1, 1) \times \mathbb{R}$ given by the "jump" set $J_M := J_M^- \cup J_M^+$, where

$$J_M^- := (-\infty, -1] \times (-\infty, -M)$$

$$\text{and}\quad J_M^+ := [1, +\infty) \times (-\infty, M).$$

Then, if M is large enough, the minimal set E_M sticks at the boundary of the slab:

Theorem 5.3.2 *There exist $M_o > 0$, $C_o > 0$, depending on s, such that if $M \geq M_o$ then*

$$[-1, 1) \times [C_o M^{\frac{1+2s}{2+2s}}, M] \subseteq E_M^c \tag{5.30}$$

$$and\quad (-1, 1] \times [-M, -C_o M^{\frac{1+2s}{2+2s}}] \subseteq E_M. \tag{5.31}$$

The situation of Theorem 5.3.2 is described in Fig. 5.11. We mention that the "strange" exponent $\frac{1+2s}{2+2s}$ in (5.30) and (5.31) is optimal.

For the detailed proof of Theorems 5.3.1 and 5.3.2, and other results on the boundary behavior of nonlocal minimal surfaces, see [63]. Here, we limit ourselves to give some heuristic motivation and a sketch of the proofs.

As a motivation for the (somehow unexpected) stickiness property at the boundary, one may look at Fig. 5.10 and argue like this. In the classical case, corresponding to $s = 1/2$, independently on the width δ, the set of minimal perimeter in B_1 will always be the half-ball $B_1 \cap \{x_n < 0\}$.

Now let us take $s < 1/2$. Then, the half-ball $B_1 \cap \{x_n < 0\}$ cannot be an s-minimal set, since the nonlocal mean curvature, for instance, at the origin cannot vanish. Indeed, the origin "sees" the complement of the set in a larger proportion than the set itself. More precisely, in B_1 (or even in $B_{1+\delta}$) the proportion of the set is the same as the one of the complement, but outside $B_{1+\delta}$ the complement of the set is dominant. Therefore, to "compensate" this lack of balance, the s-minimal set for $s < 1/2$ has to bend a bit. Likely, the s-minimal set in this case will have the tendency to become slightly convex at the origin, so that, at least nearby, it sees a proportion of the set which is larger than the proportion of the complement (we recall that, in any case, the proportion of the complement will be larger at infinity, so the set needs to compensate at least near the origin). But when δ is very small, it turns out that this compensation is not sufficient to obtain the desired balance between the set and its complement: therefore, the set has to "stick" to the half-sphere, in order to drop its constrain to satisfy a vanishing nonlocal mean curvature equation.

Of course some quantitative estimates are needed to make this argument work, so we describe the sketch of the rigorous proof of Theorem 5.3.1 as follows.

Proof (Sketch of the proof of Theorem 5.3.1) First of all, one checks that for any fixed $\eta > 0$, if $\delta > 0$ is small enough, we have that the interaction between B_1 and $B_{1+\delta} \setminus B_1$ is smaller than η. In particular, by comparing with a competitor that is empty in B_1, by minimality we obtain that

$$\mathrm{Per}_s(E_\delta, B_1) \leq \eta, \tag{5.32}$$

where we have denoted by E_δ the s-minimal set in B_1 that coincides with K_δ outside B_1.

Then, one checks that

$$\text{the boundary of } E_\delta \text{ can only lie in a small neighborhood of } \partial B_1 \tag{5.33}$$

if δ is sufficiently small.

Indeed, if, by contradiction, there were points of ∂E_δ at distance larger than ϵ from ∂B_1, then one could find two balls of radius comparable to ϵ, whose centers lie at distance larger than $\epsilon/2$ from ∂B_1 and at mutual distance smaller than ϵ, and such that one ball is entirely contained in $B_1 \cap E_\delta$ and the other ball is entirely contained in $B_1 \setminus E_\delta$ (this is due to a Clean Ball Condition, see Corollary 4.3 in [28]). As a consequence, $\mathrm{Per}_s(E_\delta, B_1)$ is bounded from below by the interaction of these two balls, which is at least of the order of ϵ^{n-2s}. Then, we obtain a contradiction with (5.32) (by choosing η much smaller than ϵ^{n-2s}, and taking δ sufficiently small).

This proves (5.33). From this, it follows that

$$\text{the whole set } E_\delta \text{ must lie in a small neighborhood of } \partial B_1. \tag{5.34}$$

Indeed, if this were not so, by (5.33) the set E_δ must contain a ball of radius, say $1/2$. Hence, $\text{Per}_s(E_\delta, B_1)$ is bounded from below by the interaction of this ball against $\{x_n > 0\} \setminus B_1$, which would produce a contribution of order one, which is in contradiction with (5.32).

Having proved (5.34), one can use it to complete the proof of Theorem 5.3.1 employing a geometric argument. Namely, one considers the ball B_ρ, which is outside E_δ for small $\rho > 0$, in virtue of (5.34), and then enlarges ρ until it touches ∂E_δ. If this contact occurs at some point $p \in B_1$, then the nonlocal mean curvature of E_δ at p must be zero. But this cannot occur (indeed, we know by (5.34) that the contribution of E_δ to the nonlocal mean curvature can only come from a small neighborhood of ∂B_1, and one can check, by estimating integrals, that this is not sufficient to compensate the outer terms in which the complement of E_δ is dominant).

As a consequence, no touching point between B_ρ and ∂E_δ can occur in B_1, which shows that E_δ is void inside B_1 and completes the proof of Theorem 5.3.1.

As for the proof of Theorem 5.3.2, the main arguments are based on sliding a ball of suitably large radius till it touches the set, with careful quantitative estimates. Some of the details are as follows (we refer to [63] for the complete arguments).

Proof (Sketch of the proof of Theorem 5.3.2) The first step is to prove a weaker form of stickiness as the one claimed in Theorem 5.3.2. Namely, one shows that

$$[-1, 1) \times [c_o M, M] \subseteq E_M^c \tag{5.35}$$

$$\text{and} \quad (-1, 1] \times [-M, -c_o M] \subseteq E_M, \tag{5.36}$$

for some $c_o \in (0, 1)$. Of course, the statements in (5.30) and (5.31) are stronger than the ones in (5.35) and (5.36) when M is large, since $\frac{1+2s}{2+2s} < 1$, but we will then obtain them later in a second step.

To prove (5.35), one takes balls of radius $c_o M$ and centered at $\{x_2 = t\}$, for any $t \in [c_o M, M]$. One slides these balls from left to right, till one touches ∂E_M. When M is large enough (and c_o small enough) this contact point cannot lie in $\{|x_1| < 1\}$. This is due to the fact that at least the sliding ball lies outside E_M, and the whole $\{x_2 > M\}$ lies outside E_M as well. As a consequence, these contact points see a proportion of E_M smaller than the proportion of the complement (it is true that the whole of $\{x_2 < -M\}$ lies inside E_M, but this contribution comes from further away than the ones just mentioned, provided that c_o is small enough). Therefore, contact points cannot satisfy a vanishing mean curvature equation and so they need to lie on the boundary of the domain (of course, careful quantitative estimates are necessary here, see [63], but we hope to have given an intuitive sketch of the computations needed).

In this way, one sees that all the portion $[-1, 1) \times [c_o M, M]$ is clean from the set E_M and so (5.35) is established (and (5.36) can be proved similarly).

Once (5.35) and (5.36) are established, one uses them to obtain the strongest form expressed in (5.30) and (5.31). For this, by (5.35) and (5.36), one has only to take care of points in $\{|x_2| \in [C_oM^{\frac{1+2s}{2+2s}}, c_oM]\}$. For these points, one can use again a sliding method, but, instead of balls, one has to use suitable surfaces obtained by appropriate portions of balls and adapt the calculations in order to evaluate all the contributions arising in this way.

The computations are not completely obvious (and once again we refer to [63] for full details), but the idea is, once again, that contact points that are in the set $\{|x_2| \in [C_oM^{\frac{1+2s}{2+2s}}, c_oM]\}$ cannot satisfy the balancing relation prescribed by the vanishing nonlocal mean curvature equation.

The stickiness property discussed above also has an interesting consequence in terms of the "geometric stability" of the flat s-minimal surfaces. For instance, rather surprisingly, the flat lines in the plane are "geometrically unstable" nonlocal minimal surfaces, in the sense that an arbitrarily small and compactly supported perturbation can produce a stickiness phenomenon at the boundary of the domain. Of course, the smaller the perturbation, the smaller the stickiness phenomenon, but it is quite relevant that such a stickiness property can occur for arbitrarily small (and "nice") perturbations. This means that s-minimal flat objects, in presence of a perturbation, may not only "bend" in the center of the domain, but rather "jump" at boundary points as well.

To state this phenomenon in a mathematical framework, one can consider, for fixed $\delta > 0$ the planar sets

$$H := \mathbb{R} \times (-\infty, 0),$$

$$F_- := (-3, -2) \times [0, \delta)$$

$$\text{and} \quad F_+ := (2, 3) \times [0, \delta).$$

One also fixes a set F which contains $H \cup F_- \cup F_+$ and denotes by E be the s-minimal set in $(-1, 1) \times \mathbb{R}$ among all the sets that coincide with F outside $(-1, 1) \times \mathbb{R}$. Then, this set E sticks at the boundary of the domain, according to the next result:

Theorem 5.3.3 *Fix $\epsilon_0 > 0$ arbitrarily small. Then, there exists $\delta_0 > 0$, possibly depending on ϵ_0, such that, for any $\delta \in (0, \delta_0]$,*

$$E \supseteq (-1, 1) \times (-\infty, \delta^{\frac{2+\epsilon_0}{1-2s}}].$$

The stickiness/instability property in Theorem 5.3.3 is depicted in Fig. 5.12. We remark that Theorem 5.3.3 gives a rather precise quantification of the size of the stickiness in terms of the size of the perturbation: namely the size of the stickiness in Theorem 5.3.3 is larger than the size of the perturbation to the power $\beta := \frac{2+\epsilon_0}{1-2s}$, for any $\epsilon_0 > 0$ arbitrarily small. Notice that $\beta \to +\infty$ as $s \to 1/2$, consistently with the fact that classical minimal surfaces do not stick at the boundary.

The proof of Theorem 5.3.3 is based on the construction of suitable auxiliary barriers (see Fig. 5.13). These barriers are used to detach a portion of the set in a neighborhood of the origin and their construction relies on some compensations

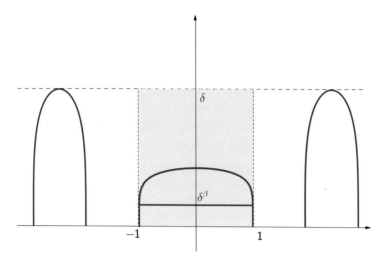

Fig. 5.12 The stickiness/instability property in Theorem 5.3.3, with $\beta := \frac{2+\epsilon_0}{1-2s}$

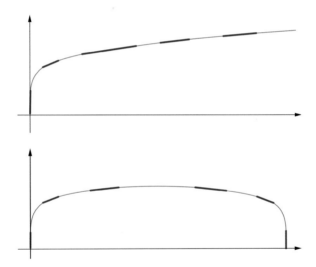

Fig. 5.13 Auxiliary barrier for the proof of Theorem 5.3.3

of nonlocal integral terms. In a sense, the building blocks of these barriers are "self-sustaining solutions" that can be seen as the geometric counterparts of the s-harmonic function x_+^s discussed in Sect. 2.4.

Indeed, roughly speaking, like the function x_+^s, these barriers "see" a proportion of the set in $\{x_1 < 0\}$ larger than what is produced by their tangent plane, but a proportion smaller than that at infinity, due to their sublinear behavior. Once again, the computations needed to check such a balancing conditions are a bit involved, and we refer to [63] for the complete details.

   To conclude this chapter, we make a remark on the connection between solutions of the fractional Allen-Cahn equation and s-minimal surfaces. Namely, a suitably scaled version of the functional in (4.9) Γ-converges to either the classical perimeter or the nonlocal perimeter functional, depending on the fractional parameter s. The Γ-convergence is a type of convergence of functionals that is compatible with the minimization of the energy, and turns out to be very useful when dealing with variational problems indexed by a parameter. This notion was introduced by De Giorgi, see e.g. [50] for details.

   In the nonlocal case, some care is needed to introduce the "right" scaling of the functional, which comes from the dilation invariance of the space coordinates and possesses a nontrivial energy in the limit. For this, one takes first the rescaled energy functional

$$ J_\varepsilon(u, \Omega) := \varepsilon^{2s} \mathscr{K}(u, \Omega) + \int_\Omega W(u)\, dx, $$

where $\mathscr{K}$ is the kinetic energy defined in (4.10). Then, one considers the functional

$$ F_\varepsilon(u, \Omega) := \begin{cases} \varepsilon^{-2s} J_\varepsilon(u, \Omega) & \text{if } s \in (0,\, 1/2), \\[2mm] |\varepsilon \log \varepsilon|^{-1} J_\varepsilon(u, \Omega) & \text{if } s = 1/2, \\[2mm] \varepsilon^{-1} J_\varepsilon(u, \Omega) & \text{if } s \in (1/2,\, 1). \end{cases} $$

The limit functional of F_ε as $\varepsilon \to 0$ depends on s. Namely, when $s \in (0, 1/2)$, the limit functional is (up to dimensional constants that we neglect) the fractional perimeter, i.e.

$$ F(u, \Omega) := \begin{cases} \mathrm{Per}_s(E, \Omega) & \text{if } u|_\Omega = \chi_E - \chi_{E^\mathscr{C}}, \text{ for some set } E \subset \Omega \\ +\infty & \text{otherwise.} \end{cases} \tag{5.37} $$

On the other hand, when $s \in [1/2, 1)$, the limit functional of F_ε is (again, up to normalizing constants) the classical perimeter, namely

$$ F(u, \Omega) := \begin{cases} \mathrm{Per}(E, \Omega) & \text{if } u|_\Omega = \chi_E - \chi_{E^\mathscr{C}}, \text{ for some set } E \subset \Omega \\ +\infty & \text{otherwise,} \end{cases} \tag{5.38} $$

That is, the following limit statement holds true:

Theorem 5.3.4 *Let $s \in (0, 1)$. Then, F_ε Γ-converges to F, as defined in either* (5.37) *or* (5.38), *depending on whether $s \in (0, 1/2)$ or $s \in [1/2, 1)$.*

   For precise statements and further details, see [123]. Additionally, we remark that the level sets of the minimizers of the functional in (4.9), after a homogeneous scaling in the space variables, converge locally uniformly to minimizers either of the fractional perimeter (if $s \in (0, 1/2)$) or of the classical perimeter (if $s \in [1/2, 1)$): that is, the "functional" convergence stated in Theorem 5.3.4 has also a "geometric" counterpart: for this, see Corollary 1.7 in [125].

One can also interpret Theorem 5.3.4 by saying that a nonlocal phase transition possesses two parameters, ε and s. When $\varepsilon \to 0$, the limit interface approaches a minimal surface either in the fractional case (when $s \in (0, 1/2)$) or in the classical case (when $s \in [1/2, 1)$). This bifurcation at $s = 1/2$ somehow states that for lower values of s the nonlocal phase transition possesses a nonlocal interface in the limit, but for larger values of s the limit interface is characterized only by local features (in a sense, when $s \in (0, 1/2)$ the "surface tension effect" is nonlocal, but for $s \in [1/2, 1)$ this effect localizes).

It is also interesting to compare Theorems 5.2 and 5.3.4, since the bifurcation at $s = 1/2$ detected by Theorem 5.3.4 is perfectly compatible with the limit behavior of the fractional perimeter, which reduces to the classical perimeter exactly for this value of s, as stated in Theorem 5.2.

Chapter 6
A Nonlocal Nonlinear Stationary Schrödinger Type Equation

The type of problems introduced in this chapter are connected to solitary solutions of nonlinear dispersive wave equations (such as the Benjamin-Ono equation, the Benjamin-Bona-Mahony equation and the fractional Schrödinger equation). In this chapter, only stationary equations are studied and we redirect the reader to [137, 138] for the study of evolutionary type equations.

Let $n \geq 2$ be the dimension of the reference space, $s \in (0, 1)$ be the fractional parameter, and $\epsilon > 0$ be a small parameter. We consider the so-called fractional Sobolev exponent

$$2_s^* := \begin{cases} \dfrac{2n}{n - 2s} & \text{for } n \geq 3, \text{ or } n = 2 \text{ and } s \in (0, 1/2) \\ +\infty & \text{for } n = 1 \text{ and } s \in (0, 1/2] \end{cases}$$

and introduce the following nonlocal nonlinear Schrödinger equation

$$\begin{cases} \epsilon^{2s}(-\Delta)^s u + u = u^p & \text{in } \Omega \subset \mathbb{R}^n \\ u = 0 & \text{in } \mathbb{R}^n \setminus \Omega, \end{cases} \tag{6.1}$$

in the subcritical case $p \in (1, 2_s^* - 1)$, namely when $p \in \left(1, \dfrac{n + 2s}{n - 2s}\right)$.

This equation arises in the study of the fractional Schrödinger equation when looking for standing waves. Namely, the fractional Schrödinger equation considers solutions $\Psi = \Psi(x, t) : \mathbb{R}^n \times \mathbb{R} \to \mathbb{C}$ of

$$i\hbar \partial_t \Psi = \left(\hbar^{2s}(-\Delta)^s + V\right)\Psi, \tag{6.2}$$

where $s \in (0, 1)$, $\hbar$ is the reduced Planck constant and $V = V(x, t, |\Psi|)$ is a potential. This equation is of interest in quantum mechanics (see e.g. [101] and the appendix in [46] for details and physical motivations). Roughly speaking, the

© Springer International Publishing Switzerland 2016
C. Bucur, E. Valdinoci, *Nonlocal Diffusion and Applications*, Lecture Notes
of the Unione Matematica Italiana 20, DOI 10.1007/978-3-319-28739-3_6

quantity $|\Psi(x, t)|^2 \, dx$ represents the probability density of finding a quantum particle in the space region dx and at time t.

The simplest solutions of (6.2) are the ones for which this probability density is independent of time, i.e. $|\Psi(x, t)| = u(x)$ for some $u : \mathbb{R}^n \to [0, +\infty)$. In this way, one can write Ψ as u times a phase that oscillates (very rapidly) in time: that is one may look for solutions of (6.2) of the form

$$\Psi(x, t) := u(x) \, e^{i\omega t / \hbar},$$

for some frequency $\omega \in \mathbb{R}$. Choosing $V = V(|\Psi|) = -|\Psi|^{p-1} = -u^{p-1}$, a substitution into (6.2) gives that

$$\left(\hbar^{2s}(-\Delta)^s u + \omega u - u^p \right) e^{i\omega t / \hbar} = \hbar^{2s}(-\Delta)^s \Psi - i\hbar \partial_t \Psi + V\Psi = 0,$$

which is (6.1) (with the normalization convention $\omega := 1$ and $\epsilon := \hbar$).

The goal of this chapter is to construct solutions of problem (6.1) that concentrate at interior points of the domain Ω for sufficiently small values of ϵ. We perform a blow-up of the domain, defined as

$$\Omega_\epsilon := \frac{1}{\epsilon}\Omega = \left\{ \frac{x}{\epsilon}, x \in \Omega \right\}.$$

We can also rescale the solution of (6.1) on Ω_ϵ,

$$u_\epsilon(x) = u(\epsilon x).$$

The problem (6.1) for u_ϵ then reads

$$\begin{cases} (-\Delta)^s u + u = u^p & \text{in } \Omega_\epsilon \\ u = 0 & \text{in } \mathbb{R}^n \setminus \Omega_\epsilon. \end{cases} \tag{6.3}$$

When $\epsilon \to 0$, the domain Ω_ϵ invades the whole of the space. Therefore, it is also natural to consider (as a first approximation) the equation on the entire space

$$(-\Delta)^s u + u = u^p \text{ in } \mathbb{R}^n. \tag{6.4}$$

The first result that we need is that there exists an entire positive radial least energy solution $w \in H^s(\mathbb{R}^n)$ of (6.4), called the *ground state solution*. Here follow some relevant results on this. The interested reader can find their proofs in [83].

1. The ground state solution $w \in H^s(\mathbb{R}^n)$ is unique up to translations.
2. The ground state solution $w \in H^s(\mathbb{R}^n)$ is nondegenerate, i.e., the derivatives $D_i w$ are solutions to the linearized equation

$$(-\Delta)^s Z + Z = pZ^{p-1}. \tag{6.5}$$

3. The ground state solution $w \in H^s(\mathbb{R}^n)$ decays polynomially at infinity, namely there exist two constants $\alpha, \beta > 0$ such that

$$\alpha |x|^{-(n+2s)} \leq u(x) \leq \beta |x|^{-(n+2s)}.$$

Unlike the fractional case, we remark that for the (classical) Laplacian, at infinity the ground state solution decays exponentially fast. We also refer to [82] for the one-dimensional case.

The main theorem of this chapter establishes the existence of a solution that concentrates at interior points of the domain for sufficiently small values of ϵ. This concentration phenomena is written in terms of the ground state solution w. Namely, the first approximation for the solution is exactly the ground state w, scaled and concentrated at an appropriate point ξ of the domain. More precisely, we have:

Theorem 6.1 *If ϵ is sufficiently small, there exist a point $\xi \in \Omega$ and a solution U_ϵ of the problem* (6.1) *such that*

$$\left| U_\epsilon(x) - w\left(\frac{x - \xi}{\epsilon}\right) \right| \leq C\epsilon^{n+2s},$$

and $\mathrm{dist}(\xi, \partial\Omega) \geq \delta > 0$. *Here, C and δ are constants independent of ϵ or Ω, and the function w is the ground state solution of problem* (6.4).

The concentration point ξ in Theorem 6.1 is influenced by the global geometry of the domain. On the one hand, when $s = 1$, the point ξ is the one that maximizes the distance from the boundary. On the other hand, when $s \in (0, 1)$, such simple characterization of ξ does not hold anymore: in this case, ξ turns out to be asymptotically the maximum of a (complicated, but rather explicit) nonlocal functional: see [46] for more details.

We state here the basic idea of the proof of Theorem 6.1 (we refer again to [46] for more details).

Proof (Sketch of the proof of Theorem 6.1) In this proof, we make use of the Lyapunov-Schmidt procedure. Namely, rather than looking for the solution in an infinite-dimensional functional space, one decomposes the problem into two orthogonal subproblems. One of these problem is still infinite-dimensional, but it has the advantage to bifurcate from a known object (in this case, a translation of the ground state). Solving this auxiliary subproblem does not provide a true solution of the original problem, since a leftover in the orthogonal direction may remain. To kill this remainder term, one solves a second subproblem, which turns out to be finite-dimensional (in our case, this subproblem is set in $\mathbb{R}^n$, which corresponds to the action of the translations on the ground state).

A structural advantage of the problem considered lies in its variational structure. Indeed, Eq. (6.3) is the Euler-Lagrange equation of the energy functional

$$I_\epsilon(u) = \frac{1}{2} \int_{\Omega_\epsilon} \left((-\Delta)^s u(x) + u(x) \right) u(x) \, dx - \frac{1}{p+1} \int_{\Omega_\epsilon} u^{p+1}(x) \, dx \qquad (6.6)$$

for any $u \in H_0^s(\Omega_\epsilon) := \{u \in H^s(\mathbb{R}^n) \text{ s.t. } u = 0 \text{ a.e. in } \mathbb{R}^n \setminus \Omega_\epsilon\}$. Therefore, the problem reduces to finding critical points of I_ϵ.

To this goal, we consider the *ground state* solution w and for any $\xi \in \mathbb{R}^n$ we let $w_\xi := w(x - \xi)$. For a given $\xi \in \Omega_\epsilon$ a first approximation $\bar{u}_\xi$ for the solution of problem (6.3) can be taken as the solution of the linear problem

$$\begin{cases} (-\Delta)^s \bar{u}_\xi + \bar{u}_\xi = w_\xi^p & \text{in } \Omega_\epsilon, \\ \bar{u}_\xi = 0 & \text{in } \mathbb{R}^n \setminus \Omega_\epsilon. \end{cases} \tag{6.7}$$

The actual solution will be obtained as a small perturbation of $\bar{u}_\xi$ for a suitable point $\xi = \xi(\epsilon)$. We define the operator $\mathscr{L} := (-\Delta)^s + I$, where I is the identity and we notice that $\mathscr{L}$ has a unique fundamental solution that solves

$$\mathscr{L}\Gamma = \delta_0 \quad \text{in } \mathbb{R}^n.$$

The Green function G_ϵ of the operator $\mathscr{L}$ in Ω_ϵ satisfies

$$\begin{cases} \mathscr{L}G_\epsilon(x, y) = \delta_y(x) & \text{if } x \in \Omega_\epsilon, \\ G_\epsilon(x, y) = 0 & \text{if } x \in \mathbb{R}^n \setminus \Omega_\epsilon. \end{cases} \tag{6.8}$$

It is convenient to introduce the regular part of G_ϵ, which is often called the Robin function. This function is defined by

$$H_\epsilon(x, y) := \Gamma(x - y) - G_\epsilon(x, y) \tag{6.9}$$

and it satisfies, for a fixed $y \in \mathbb{R}^n$,

$$\begin{cases} \mathscr{L}H_\epsilon(x, y) = 0 & \text{if } x \in \Omega_\epsilon, \\ H_\epsilon(x, y) = \Gamma(x - y) & \text{if } x \in \mathbb{R}^n \setminus \Omega_\epsilon. \end{cases} \tag{6.10}$$

Then

$$\bar{u}_\xi(x) = \int_{\Omega_\epsilon} \bar{u}_\xi(y) \delta_0(x - y) \, dy,$$

and by (6.8)

$$\bar{u}_\xi(x) = \int_{\Omega_\epsilon} \bar{u}_\xi(y) \mathscr{L}G_\epsilon(x, y) \, dy.$$

The operator $\mathcal{L}$ is self-adjoint and thanks to the above identity and to Eq. (6.7) it follows that

$$\bar{u}_\xi(x) = \int_{\Omega_\epsilon} \mathcal{L}\bar{u}_\xi(y)G_\epsilon(x, y)\, dy$$

$$= \int_{\Omega_\epsilon} w_\xi^p(y)G_\epsilon(x, y)\, dy.$$

So, we use (6.9) and we obtain that

$$\bar{u}_\xi(x) = \int_{\Omega_\epsilon} w_\xi^p(y)\Gamma(x - y)\, dy - \int_{\Omega_\epsilon} w_\xi^p(y)H_\epsilon(x, y)\, dy.$$

Now we notice that, since w_ξ is solution of (6.4) and Γ is the fundamental solution of $\mathcal{L}$, we have that

$$\int_{\mathbb{R}^n} w_\xi^p(y)\Gamma(x - y)\, dy = \int_{\mathbb{R}^n} \mathcal{L}w_\xi(y)\Gamma(x - y)\, dy$$

$$= \int_{\mathbb{R}^n} w_\xi(y)\mathcal{L}\Gamma(x - y)\, dy$$

$$= w_\xi(x).$$

Therefore we have obtained that

$$\bar{u}_\xi(x) = w_\xi(x) - \int_{\mathbb{R}^n \setminus \Omega_\epsilon} w_\xi^p(y)\Gamma(x - y)\, dy - \int_{\Omega_\epsilon} w_\xi^p(y)H_\epsilon(x, y)\, dy. \qquad (6.11)$$

Now we can insert (6.11) into the energy functional (6.6) and expand the errors in powers of ϵ. For $\mathrm{dist}(\xi, \partial\Omega_\epsilon) \geq \dfrac{\delta}{\epsilon}$ with δ fixed and small, the energy of $\bar{u}_\xi$ is a perturbation of the energy of the ground state w and one finds (see Theorem 4.1 in [46]) that

$$I_\epsilon(\bar{u}_\xi) = I(w) + \frac{1}{2}\mathcal{H}_\epsilon(\xi) + \mathcal{O}(\epsilon^{n+4s}), \qquad (6.12)$$

where

$$\mathcal{H}_\epsilon(\xi) := \int_{\Omega_\epsilon} \int_{\Omega_\epsilon} H_\epsilon(x, y)w_\xi^p(x)w_\xi^p(y)\, dx\, dy$$

and I is the energy computed on the whole space $\mathbb{R}^n$, namely

$$I(u) = \frac{1}{2}\int_{\mathbb{R}^n} \left((-\Delta)^s u(x) + u(x)\right)u(x)\, dx - \frac{1}{p+1}\int_{\mathbb{R}^n} u^{p+1}(x)\, dx. \qquad (6.13)$$

In particular, $I_\epsilon(\overline{u}_\xi)$ agrees with a constant (the term $I(w)$), plus a functional over a finite-dimensional space (the term $\mathscr{H}_\epsilon(\xi)$, which only depends on $\xi \in \mathbb{R}^n$), plus a small error.

We remark that the solution $\overline{u}_\xi$ of Eq. (6.7) which can be obtained from (6.11) does not provide a solution for the original problem (6.3) (indeed, it only solves (6.7)): for this, we look for solutions u_ξ of (6.3) as perturbations of $\overline{u}_\xi$, in the form

$$u_\xi := \overline{u}_\xi + \psi. \qquad (6.14)$$

The perturbation functions ψ are considered among those vanishing outside Ω_ϵ and orthogonal to the space $\mathscr{Z} = \mathrm{Span}(Z_1, \ldots, Z_n)$, where $Z_i = \dfrac{\partial w_\xi}{\partial x_i}$ are solutions of the linearized Eq. (6.5). This procedure somehow "removes the degeneracy", namely we look for the corrector ψ in a set where the linearized operator is invertible. This makes it possible, fixed any $\xi \in \mathbb{R}^n$, to find $\psi = \psi_\xi$ such that the function u_ξ, as defined in (6.14) solves the equation

$$(-\Delta)^s u_\xi + u_\xi = u_\xi^p + \sum_{i=1}^n c_i Z_i \text{ in } \Omega_\epsilon. \qquad (6.15)$$

That is, u_ξ is solution of the original Eq. (6.3), up to an error that lies in the tangent space of the translations (this error is exactly the price that we pay in order to solve the corrector equation for ψ on the orthogonal of the kernel, where the operator is nondegenerate). As a matter of fact (see Theorem 7.6 in [46] for details) one can see that the corrector $\psi = \psi_\xi$ is of order ϵ^{n+2s}. Therefore, one can compute $I_\epsilon(u_\xi) = I_\epsilon(\overline{u}_\xi + \psi_\xi)$ as a higher order perturbation of $I_\epsilon(\overline{u}_\xi)$. From (6.12), one obtains that

$$I_\epsilon(u_\xi) = I(w) + \frac{1}{2}\mathscr{H}_\epsilon(\xi) + \mathcal{O}(\epsilon^{n+4s}), \qquad (6.16)$$

see Theorem 7.17 in [46] for details.

Since this energy expansion now depends only on $\xi \in \mathbb{R}^n$, it is convenient to define the operator $J_\epsilon : \Omega_\epsilon \to \mathbb{R}$ as

$$J_\epsilon(\xi) := I_\epsilon(u_\xi).$$

This functional is often called the reduced energy functional. From (6.16), we conclude that

$$J_\epsilon(\xi) = I(w) + \frac{1}{2}\mathscr{H}_\epsilon(\xi) + \mathcal{O}(\epsilon^{n+4s}). \qquad (6.17)$$

The reduced energy J plays an important role in this framework since critical points of J correspond to true solutions of the original Eq. (6.3). More precisely (see

Lemma 7.16 in [46]) one has that $c_i = 0$ for all $i = 1,\ldots,n$ in (6.15) if and only if

$$\frac{\partial J_\epsilon}{\partial \xi}(\xi) = 0. \tag{6.18}$$

In other words, when ϵ approaches 0, to find concentration points, it is enough to find critical points of J, which is a finite-dimensional problem. Also, critical points for J come from critical points of $\mathcal{H}_\epsilon$, up to higher orders, thanks to (6.17). The issue is thus to prove that $\mathcal{H}_\epsilon$ does possess critical points and that these critical points survive after the small error of size ϵ^{n+4s}: in fact, we show that $\mathcal{H}_\epsilon$ possesses a minimum, which is stable for perturbations. For this, one needs a bound for the Robin function H_ϵ from above and below. To this goal, one builds a barrier function β_ξ defined for $\xi \in \Omega_\epsilon$ and $x \in \mathbb{R}^n$ as

$$\beta_\xi(x) := \int_{\mathbb{R}^n \setminus \Omega_\epsilon} \Gamma(z - \xi)\Gamma(x - z)\, dz.$$

Using this function in combination with suitable maximum principles, one obtains the existence of a constant $c \in (0, 1)$ such that

$$cH_\epsilon(x, \xi) \leq \beta_\xi(x) \leq c^{-1}H_\epsilon(x, \xi),$$

for any $x \in \mathbb{R}^n$ and any $\xi \in \Omega_\epsilon$ with $\mathrm{dist}(\xi, \partial\Omega_\epsilon) > 1$, see Lemma 2.1 in [46]. From this it follows that

$$\mathcal{H}_\epsilon(\xi) \simeq d^{-(n+4s)}, \tag{6.19}$$

for all points $\xi \in \Omega_\epsilon$ such that $d \in [5, \delta/\epsilon]$. So, one considers the domain $\Omega_{\epsilon,\delta}$ of the points of Ω_ϵ that lie at distance more than δ/ϵ from the boundary of Ω_ϵ. By (6.19), we have that

$$\mathcal{H}_\epsilon(\xi) \simeq \frac{\epsilon^{n+4s}}{\delta^{n+4s}} \quad \text{for any } \xi \in \partial\Omega_{\epsilon,\delta}. \tag{6.20}$$

Also, up to a translation, we may suppose that $0 \in \Omega$. Thus, $0 \in \Omega_\epsilon$ and its distance from $\partial\Omega_\epsilon$ is of order $1/\epsilon$ (independently of δ). In particular, if δ is small enough, we have that 0 lies in the interior of $\Omega_{\epsilon,\delta}$, and (6.19) gives that

$$\mathcal{H}_\epsilon(0) \simeq \epsilon^{n+4s}.$$

By comparing this with (6.20), we see that $\mathcal{H}_\epsilon$ has an interior minimum in $\Omega_{\epsilon,\delta}$. The value attained at this minimum is of order ϵ^{n+4s}, and the values attained at the boundary of $\Omega_{\epsilon,\delta}$ are of order $\delta^{-n-4s}\epsilon^{n+4s}$, which is much larger than ϵ^{n+4s}, if δ is

Fig. 6.1 Geometric
interpretation

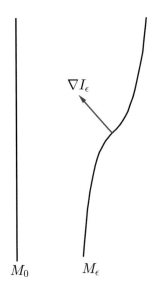

$$\nabla I_\epsilon$$

$$M_0 \qquad M_\epsilon$$

small enough. This says that the interior minimum of $\mathscr{H}_\epsilon$ in $\Omega_{\epsilon,\delta}$ is nondegenerate
and it survives to any perturbation of order ϵ^{n+4s}, if δ is small enough.

This and (6.17) imply that J has also an interior minimum at some point ξ in $\Omega_{\epsilon,\delta}$.
By construction, this point ξ satisfies (6.18), and so this completes the proof of
Theorem 6.1.

The variational argument in the proof above (see in particular (6.18)) has a
classical and neat geometric interpretation. Namely, the "unperturbed" functional
(i.e. the one with $\epsilon = 0$) has a very degenerate geometry, since it has a whole
manifold of minimizers with the same energy: this manifold corresponds to the
translation of the ground state w, namely it is of the form $M_0 := \{w_\xi, \ \xi \in \mathbb{R}^n\}$
and, therefore, it can be identified with $\mathbb{R}^n$.

For topological arguments, this degenerate picture may constitute a serious
obstacle to the existence of critical points for the "perturbed" functional (i.e. the one
with $\epsilon \neq 0$). As an obvious example, the reader may think of the function of two
variables $f_\epsilon : \mathbb{R}^2 \to \mathbb{R}$ given by $f_\epsilon(x,y) := x^2 + \epsilon y$. When $\epsilon = 0$, this function attains
its minimum along the manifold $\{x = 0\}$, but all the critical points on this manifold
are "destroyed" by the perturbation when $\epsilon \neq 0$ (indeed $\nabla f_\epsilon(x,y) = (2x, \epsilon)$ never
vanishes).

In the situation described in the proof of Theorem 6.1, this pathology does not
occur, thanks to the nondegeneracy provided in [83]. Indeed, by the nondegeneracy
of the unperturbed critical manifold, when $\epsilon \neq 0$ one can construct a manifold,
diffeomorphic to the original one (in our case of the form $M_\epsilon := \{\bar{u}_\xi + \psi(\xi), \ \xi \in \mathbb{R}^n\}$), that enjoys the special feature of "almost annihilating" the gradient of the
functional, up to vectors parallel to the original manifold M_0 (this is the meaning of
formula (6.15)) (see also Fig. 6.1).

Then, if one finds a minimum of the functional constrained to M_ϵ, the theory of
Lagrange multipliers (at least in finite dimension) would suggest that the gradient
is normal to M_ϵ. That is, the gradient of the functional is, simultaneously, parallel
to M_0 and orthogonal to M_ϵ. But since M_ϵ is diffeomorphically close to M_0, the only
vector with this property is the null vector, hence this argument provides the desired
critical point.

We also recall that the fractional Schrödinger equation is related to a nonlocal
canonical quantization, which in turn produces a nonlocal Uncertainty Principle.
In the classical setting, one considers the momentum/position operators, which are
defined in $\mathbb{R}^n$, by

$$P_k := -i\hbar\partial_k \quad \text{and} \quad Q_k := x_k \tag{6.21}$$

for $k \in \{1,\dots,n\}$. Then, the Uncertainty Principle states that the operators $P = (P_1,\dots,P_n)$ and $Q = (Q_1,\dots,Q_n)$ do not commute (which makes it practically
impossible to measure simultaneously both momentum and position). Indeed, in
this case a simple computation shows that

$$[Q,P] := \sum_{k=1}^{n}[Q_k,P_k] = i\hbar n. \tag{6.22}$$

The nonlocal analogue of this quantization may be formulated by introducing a
nonlocal momentum, i.e. by replacing the operators in (6.21) by

$$P_k^s := -i\hbar^s\partial_k(-\Delta)^{\frac{s-1}{2}} \quad \text{and} \quad Q_k := x_k. \tag{6.23}$$

In this case, using that the Fourier transform of the product is the convolution of the
Fourier transforms, one has that

$$
\begin{aligned}
(\hat{x}_k * g)(\xi) &= \mathscr{F}\big(x_k\mathscr{F}^{-1}g(x)\big)(\xi) \\
&= \int_{\mathbb{R}^n} dx \int_{\mathbb{R}^n} dy \, e^{2\pi i x\cdot(y-\xi)}x_k g(y) \\
&= \frac{1}{2\pi i}\int_{\mathbb{R}^n} dx \int_{\mathbb{R}^n} dy \, \partial_{y_k}e^{2\pi i x\cdot(y-\xi)}g(y) \\
&= \frac{i}{2\pi}\int_{\mathbb{R}^n} dx \int_{\mathbb{R}^n} dy \, e^{2\pi i x\cdot(y-\xi)}\partial_k g(y) \\
&= \frac{i}{2\pi}\int_{\mathbb{R}^n} dx \, e^{-2\pi i x\cdot\xi}\mathscr{F}^{-1}(\partial_k g)(x) \\
&= \frac{i}{2\pi}\mathscr{F}\big(\mathscr{F}^{-1}(\partial_k g)\big)(\xi) \\
&= \frac{i}{2\pi}\partial_k g(\xi),
\end{aligned}
\tag{6.24}
$$

for any test function g. In addition,

$$\mathscr{F}(P_k^s f) = (2\pi)^s \hbar^s \xi_k |\xi|^{s-1} \hat{f}.$$

Therefore, given any test function ψ, using this with $f := \psi$ and $f := x_k \psi$, and also (6.24) with $g := \mathscr{F}(P_k^s \psi)$ and $g := \hat{\psi}$, we obtain that

$$\mathscr{F}\left(Q_k P_k^s \psi(x) - P_k^s Q_k \psi(x)\right)$$
$$= \mathscr{F}\left(x_k P_k^s \psi(x) - P_k^s(x_k \psi(x))\right)$$
$$= \hat{x}_k * \mathscr{F}(P_k^s \psi(x)) - \mathscr{F}(P_k^s(x_k \psi(x)))$$
$$= \frac{i}{2\pi} \partial_k \mathscr{F}(P_k^s \psi)(\xi) - (2\pi)^s \hbar^s \xi_k |\xi|^{s-1} \mathscr{F}(x_k \psi(x))(\xi)$$
$$= (2\pi)^{s-1} i\hbar^s \partial_k\left(\xi_k |\xi|^{s-1} \hat{\psi}(\xi)\right) - (2\pi)^s \hbar^s \xi_k |\xi|^{s-1} \hat{x}_k * \hat{\psi}(\xi)$$
$$= (2\pi)^{s-1} i\hbar^s \partial_k\left(\xi_k |\xi|^{s-1} \hat{\psi}(\xi)\right) - (2\pi)^{s-1} i\hbar^s \xi_k |\xi|^{s-1} \partial_k \hat{\psi}(\xi)$$
$$= (2\pi)^{s-1} i\hbar^s \partial_k\left(\xi_k |\xi|^{s-1}\right) \hat{\psi}(\xi)$$
$$= (2\pi)^{s-1} i\hbar^s \left(|\xi|^{s-1} + (s-1)\xi_k^2 |\xi|^{s-3}\right) \hat{\psi}(\xi).$$

Consequently, by summing up,

$$\mathscr{F}\left([Q, P^s]\psi\right) = (2\pi)^{s-1} i\hbar^s |\xi|^{s-1} (n+s-1) \hat{\psi}(\xi).$$

So, by taking the anti-transform,

$$[Q, P^s]\psi = i\hbar^s (n+s-1) \mathscr{F}^{-1}\left((2\pi|\xi|)^{s-1} \hat{\psi}\right)$$
$$= i\hbar^s (n+s-1)(-\Delta)^{\frac{s-1}{2}} \psi. \tag{6.25}$$

Notice that, as $s \to 1$, this formula reduces to the the classical Heisenberg Uncertainty Principle in (6.22).

6.1   From the Nonlocal Uncertainty Principle to a Fractional Weighted Inequality

Now we point out a simple consequence of the Uncertainty Principle in formula (6.25), which can be seen as a fractional Sobolev inequality in weighted spaces. The result (which boils down to known formulas as $s \to 1$) is the following:

Proposition 6.1 *For any* $u \in \mathscr{S}(\mathbb{R}^n)$, *we have that*

$$\left\| (-\Delta)^{\frac{s-1}{4}} u \right\|_{L^2(\mathbb{R}^n)}^2 \leq \frac{2}{n+s-1} \left\| |x| u \right\|_{L^2(\mathbb{R}^n)} \left\| \nabla (-\Delta)^{\frac{s-1}{2}} u \right\|_{L^2(\mathbb{R}^n)}.$$

Proof The proof is a general argument in operator theory. Indeed, suppose that there are two operators S and A, acting on a space with a scalar Hermitian product. Assume that S is self-adjoint and A is anti-self-adjoint, i.e.

$$\langle u, Su \rangle = \langle Su, u \rangle \quad \text{and} \quad \langle u, Au \rangle = -\langle Au, u \rangle,$$

for any u in the space. Then, for any $\lambda \in \mathbb{R}$,

$$\|(A + \lambda S)u\|^2 = \|Au\|^2 + \lambda^2 \|Su\|^2 + \lambda \Big(\langle Au, Su \rangle + \langle Su, Au \rangle \Big)$$

$$= \|Au\|^2 + \lambda^2 \|Su\|^2 + \lambda \langle (SA - AS)u, u \rangle.$$

Now we apply this identity in the space $C_0^\infty(\mathbb{R}^n) \subset L^2(\mathbb{R}^n)$, taking $S := Q_k = x_k$ and $A := iP_k^s = \hbar^s \partial_k (-\Delta)^{\frac{s-1}{2}}$ (recall (6.23) and notice that iP_k^s is anti-self-adjoint, thanks to the integration by parts formula). In this way, and using (6.25), we obtain that

$$0 \leq \sum_{k=1}^n \|(iP_k^s + \lambda Q_k)u\|_{L^2(\mathbb{R}^n)}^2$$

$$= \sum_{k=1}^n \left[\|P_k^s u\|_{L^2(\mathbb{R}^n)}^2 + \lambda^2 \|Q_k u\|_{L^2(\mathbb{R}^n)}^2 + i\lambda \langle [Q_k, P_k^s] u, u \rangle_{L^2(\mathbb{R}^n)} \right]$$

$$= \hbar^{2s} \left\| \nabla (-\Delta)^{\frac{s-1}{2}} u \right\|_{L^2(\mathbb{R}^n)}^2 + \lambda^2 \left\| |x| u \right\|_{L^2(\mathbb{R}^n)}^2$$

$$+ i^2 \lambda (n+s-1) \hbar^s \langle (-\Delta)^{\frac{s-1}{2}} u, u \rangle_{L^2(\mathbb{R}^n)}$$

$$= \hbar^{2s} \left\| \nabla (-\Delta)^{\frac{s-1}{2}} u \right\|_{L^2(\mathbb{R}^n)}^2 + \lambda^2 \left\| |x| u \right\|_{L^2(\mathbb{R}^n)}^2$$

$$- \lambda (n+s-1) \hbar^s \left\| (-\Delta)^{\frac{s-1}{4}} u \right\|_{L^2(\mathbb{R}^n)}^2.$$

Now, if $u \neq 0$, we can optimize this identity by choosing

$$\lambda := \frac{(n+s-1) \hbar^s \left\| (-\Delta)^{\frac{s-1}{4}} u \right\|_{L^2(\mathbb{R}^n)}^2}{2 \left\| |x| u \right\|_{L^2(\mathbb{R}^n)}^2}$$

and we obtain that

$$0 \leq \hbar^{2s} \left\| \nabla(-\Delta)^{\frac{s-1}{2}} u \right\|_{L^2(\mathbb{R}^n)}^2 - \frac{(n+s-1)^2 \, \hbar^{2s} \left\| (-\Delta)^{\frac{s-1}{4}} u \right\|_{L^2(\mathbb{R}^n)}^4}{4 \left\| |x| u \right\|_{L^2(\mathbb{R}^n)}^2},$$

which gives the desired result.

Appendix A
Alternative Proofs of Some Results

We present in this Appendix alternative proofs of some results already introduced in this book. In the first section, we give two different proofs that the function x_+^s is s-harmonic on the positive half-line $\mathbb{R}_+$. We then alternatively compute some constants related to the fractional Laplacian.

A.1 Another Proof of Theorem 2.4.1

Here we present a different proof of Theorem 2.4.1, based on the Fourier transforms of homogeneous distributions. This proof is the outcome of a pleasant discussion with Alexander Nazarov.

Proof (Alternative proof of Theorem 2.4.1) We are going to use the Fourier transform of $|x|^q$ in the sense of distribution, with $q \in \mathbb{C} \setminus \mathbb{Z}$. Namely (see e.g. Lemma 2.23 on page 38 of [96])

$$\mathscr{F}(|x|^q) = C_q\, |\xi|^{-1-q}, \tag{A.1}$$

with

$$C_q := -2(2\pi)^{-q-1}\Gamma(1+q) \sin \frac{\pi q}{2}. \tag{A.2}$$

We remark that the original value of the constant C_q in [96] is here multiplied by a $(2\pi)^{-q-1}$ term, in order to be consistent with the Fourier normalization that we have introduced. We observe that the function $\mathbb{R} \ni x \mapsto |x|^q$ is locally integrable only when $q > -1$, so it naturally induces a distribution only in this range of the parameter q (and, similarly, the function $\mathbb{R} \ni \xi \mapsto |\xi|^{-1-q}$ is locally integrable only when $q < 0$): therefore, to make sense of the formulas above in a wider range of parameters q it is necessary to use analytic continuation and a special procedure that

© Springer International Publishing Switzerland 2016
C. Bucur, E. Valdinoci, *Nonlocal Diffusion and Applications*, Lecture Notes
of the Unione Matematica Italiana 20, DOI 10.1007/978-3-319-28739-3

is called regularization: see e.g. page 36 in [96] (as a matter of fact, we will do a procedure of this type explicitly in a particular case in (A.11)). Since $\mathbb{R} \ni x \mapsto |x|^q$ is even, we can write (A.1) also as

$$\mathcal{F}^{-1}(|\xi|^q) = C_q\,|x|^{-1-q}. \tag{A.3}$$

We observe that, by elementary trigonometry,

$$\sin\frac{\pi(s+1)}{2} = -\sin\frac{\pi(s-1)}{2} \quad\text{and}\quad \sin\frac{\pi(s-2)}{2} = \sin\frac{\pi s}{2}.$$

Moreover,

$$\Gamma(2+s) = (1+s)\Gamma(1+s) \quad\text{and}\quad \Gamma(s) = (s-1)\Gamma(s-1).$$

Hence, by (A.2),

$$\frac{1-s}{1+s}\cdot C_{s+1}\,C_{s-2} = \frac{1-s}{1+s}\cdot 4(2\pi)^{-2s-1}\Gamma(2+s)\,\Gamma(s-1)\sin\frac{\pi(s+1)}{2}\,\sin\frac{\pi(s-2)}{2}$$

$$= 4\Gamma(1+s)\,\Gamma(s)\sin\frac{\pi(s-1)}{2}\,\sin\frac{\pi s}{2} = C_s\,C_{s-1}. \tag{A.4}$$

Moreover,

$$|x|^s + \frac{1}{s+1}\partial_x|x|^{s+1} = 2x_+^s.$$

So, taking the Fourier transform and using (A.1) with $q := s$ and $q := s+1$, we obtain that

$$2\mathcal{F}(x_+^s) = \mathcal{F}(|x|^s) + \frac{1}{s+1}\mathcal{F}\left(\partial_x|x|^{s+1}\right)$$

$$= \mathcal{F}(|x|^s) + \frac{2\pi i\xi}{s+1}\mathcal{F}(|x|^{s+1})$$

$$= C_s\,|\xi|^{-1-s} + \frac{2\pi i\xi}{s+1}C_{s+1}\,|\xi|^{-2-s}.$$

As a consequence,

$$2|\xi|^{2s}\mathcal{F}(x_+^s) = C_s\,|\xi|^{-1+s} + \frac{2\pi i\xi}{s+1}C_{s+1}\,|\xi|^{-2+s}.$$

Hence, recalling (6.24),

$$2\mathscr{F}^{-1}\Big(|\xi|^{2s}\mathscr{F}(x_+^s)\Big) = C_s\,\mathscr{F}^{-1}(|\xi|^{-1+s}) + \frac{2\pi C_{s+1}\,i}{s+1}\mathscr{F}^{-1}(\xi) * \mathscr{F}^{-1}(|\xi|^{-2+s})$$

$$= C_s\,\mathscr{F}^{-1}(|\xi|^{-1+s}) - \frac{2\pi C_{s+1}\,i}{s+1}\cdot\frac{i}{2\pi}\partial_x\mathscr{F}^{-1}(|\xi|^{-2+s})$$

$$= C_s\,\mathscr{F}^{-1}(|\xi|^{-1+s}) + \frac{C_{s+1}}{s+1}\partial_x\mathscr{F}^{-1}(|\xi|^{-2+s}).$$

Accordingly, exploiting (A.3) with $q := -1+s$ and $q := -2+s$,

$$2\mathscr{F}^{-1}\Big(|\xi|^{2s}\mathscr{F}(x_+^s)\Big) = C_s\,C_{s-1}\,|x|^{-s} + \frac{C_{s+1}\,C_{s-2}}{s+1}\partial_x|x|^{1-s}$$

$$= C_s\,C_{s-1}\,|x|^{-s} + \frac{1-s}{1+s}\cdot C_{s+1}\,C_{s-2}\,x\,|x|^{-1-s}.$$

So, recalling (A.4),

$$2\mathscr{F}^{-1}\Big(|\xi|^{2s}\mathscr{F}(x_+^s)\Big) = C_s\,C_{s-1}\Big(|x|^{-s} - x\,|x|^{-1-s}\Big).$$

This and (2.7) give that

$$(-\Delta)^s(x_+^s) = \tilde{C}_s\Big(|x|^{-s} - x\,|x|^{-1-s}\Big),$$

for some $\tilde{C}_s$. In particular, the quantity above vanishes when $x > 0$, thus providing an alternative proof of Theorem 2.4.1.

Yet another proof of Theorem 2.4.1 can be obtained in a rather short, but technically quite advanced way, using the Paley-Wiener theory in the distributional setting. The sketch of this proof goes as follows:

Proof (Alternative proof of Theorem 2.4.1) The function $h := x_+^s$ is homogeneous of degree s. Therefore its (distributional) Fourier transform $\mathscr{F}h$ is homogeneous of degree $-1-s$ (see Lemma 2.21 in [96]).

Moreover, h is supported in $\{x \geq 0\}$, therefore $\mathscr{F}h$ can be continued to an analytic function in $\mathbb{C}_- := \{z \in \mathbb{C} \text{ s.t. } \Im z < 0\}$ (see Theorem 2 in [129]).

Therefore, $g(\xi) := |\xi|^s\mathscr{F}h(\xi)$ is homogeneous of degree -1 and is the trace of a function that is analytic in $\mathbb{C}_-$.

In particular, for any $y < 0$, we have that

$$g(-iy) = \frac{g(-i)}{y} = \frac{c}{-iy},$$

where $c := -ig(-i)$. That is, $g(z)$ coincides with $\frac{c}{z}$ on a half-line, and then in the whole of $\mathbb{C}_-$, by analytic continuation.

That is, in the sense of distributions,

$$g(\xi) = \frac{c}{\xi - i0},$$

for any $\xi \in \mathbb{R}$.

Now we recall the Sokhotski Formula (see e.g. (3.10) in [16]), according to which

$$\frac{1}{\xi \pm i0} = \mp i\pi\delta + \text{P.V.}\frac{1}{\xi},$$

where δ is the Dirac's Delta and the identity holds in the sense of distributions. By considering this equation with the two sign choices and then summing up, we obtain that

$$\frac{1}{\xi + i0} - \frac{1}{\xi - i0} = -2i\pi\delta.$$

As a consequence

$$g(\xi) = \frac{c}{\xi - i0} = \frac{c}{\xi + i0} + 2ic\pi\delta.$$

Therefore

$$|\xi|^{2s}\mathscr{F}h(\xi) = |\xi|^{s}g(\xi) = \frac{c\,|\xi|^{s}}{\xi + i0} + 2ic\pi\,|\xi|^{s}\delta.$$

Of course, as a distribution, $|\xi|^{s}\delta = 0$, since the evaluation of $|\xi|^{s}$ at $\xi = 0$ vanishes, therefore we can write

$$\ell(\xi) := |\xi|^{2s}\mathscr{F}h(\xi) = \frac{c\,|\xi|^{s}}{\xi + i0}.$$

Accordingly, ℓ is homogeneous of degree $-1 + s$ and it is the trace of a function analytic in $\mathbb{C}_+ := \{z \in \mathbb{C} \text{ s.t. } \Im z > 0\}$.

Consequently, $\mathscr{F}^{-1}\ell$ is homogeneous of degree $-s$ (see again Lemma 2.21 in [96]), and it is supported in $\{x \leq 0\}$ (see again Theorem 2 in [129]). Since $(-\Delta)^s x_+^s$ coincides (up to multiplicative constants) with $\mathscr{F}^{-1}\ell$, we have just shown that $(-\Delta)^s x_+^s = c_o x_-^{-s}$, for some $c_o \in \mathbb{R}$, and so in particular $(-\Delta)^s x_+^s$ vanishes in $\{x > 0\}$.

A.2 Another Proof of Lemma 2.3

For completeness, we provide here an alternative proof of Lemma 2.3 that does not use the theory of the fractional Laplacian.

Proof (Alternative proof of Lemma 2.3) We first recall some basic properties of the modified Bessel functions (see e.g. [4]). First of all (see formula 9.1.10 on page 360 of [4]), we have that

$$J_s(z) := \frac{z^s}{2^s} \sum_{k=0}^{+\infty} \frac{(-1)^k z^{2k}}{2^{2k}\, k!\, \Gamma(s+k+1)} = \frac{z^s}{2^s\, \Gamma(1+s)} + \mathcal{O}(|z|^{2+s})$$

as $|z| \to 0$. Therefore (see formula 9.6.3 on page 375 of [4]),

$$I_s(z) := e^{-\frac{i\pi s}{2}} J_s(e^{\frac{i\pi}{2}} z)$$

$$= e^{-\frac{i\pi s}{2}} \left(\frac{e^{\frac{i\pi s}{2}} z^s}{2^s\, \Gamma(1+s)} + \mathcal{O}(|z|^{2+s}) \right)$$

$$= \frac{z^s}{2^s\, \Gamma(1+s)} + \mathcal{O}(|z|^{2+s}),$$

as $|z| \to 0$. Using this and formula 9.6.2 on page 375 of [4],

$$K_s(z) := \frac{\pi}{2\sin(\pi s)} \left(I_{-s}(z) - I_s(z) \right)$$

$$= \frac{\pi}{2\sin(\pi s)} \left(\frac{z^{-s}}{2^{-s}\, \Gamma(1-s)} - \frac{z^s}{2^s\, \Gamma(1+s)} + \mathcal{O}(|z|^{2-s}) \right).$$

Thus, recalling Euler's reflection formula

$$\Gamma(1-s)\, \Gamma(s) = \frac{\pi}{\sin(\pi s)},$$

and the relation $\Gamma(1+s) = s\Gamma(s)$, we obtain

$$K_s(z) = \frac{\Gamma(1-s)\, \Gamma(s)}{2} \left(\frac{z^{-s}}{2^{-s}\, \Gamma(1-s)} - \frac{z^s}{2^s\, \Gamma(1+s)} + \mathcal{O}(|z|^{2-s}) \right)$$

$$= \frac{\Gamma(s)\, z^{-s}}{2^{1-s}} - \frac{\Gamma(1-s)\, z^s}{2^{1+s} s} + \mathcal{O}(|z|^{2-s}),$$

as $|z| \to 0$. We use this and formula (3.100) in [105] (or page 6 in [112]) and get that, for any small $a > 0$,

$$
\int_0^{+\infty} \frac{\cos(2\pi t)}{(t^2 + a^2)^{s+\frac{1}{2}}} \, dt = \frac{\pi^{s+\frac{1}{2}}}{a^s \, \Gamma\left(s + \frac{1}{2}\right)} K_s(2\pi a)
$$

$$
= \frac{\pi^{s+\frac{1}{2}}}{a^s \, \Gamma\left(s + \frac{1}{2}\right)} \left[\frac{\Gamma(s)}{2\pi^s a^s} - \frac{\Gamma(1-s)\,\pi^s a^s}{2s} + \mathcal{O}(a^{2-s}) \right] \tag{A.5}
$$

$$
= \frac{\pi^{\frac{1}{2}} \Gamma(s)}{2a^{2s} \, \Gamma\left(s + \frac{1}{2}\right)} - \frac{\pi^{2s+\frac{1}{2}} \Gamma(1-s)}{2s\Gamma\left(s + \frac{1}{2}\right)} + \mathcal{O}(a^{2-2s}).
$$

Now we recall the generalized hypergeometric functions $_mF_n$ (see e.g. page 211 in [112]): as a matter of fact, we just need that for any $b, c, d > 0$,

$$
_1F_2(b; c, d; 0) = \frac{\Gamma(c)\Gamma(d)}{\Gamma(b)} \cdot \frac{\Gamma(b)}{\Gamma(c)\Gamma(d)} = 1.
$$

We also recall the Beta function relation

$$
B(\alpha, \beta) = \frac{\Gamma(\alpha)\,\Gamma(\beta)}{\Gamma(\alpha + \beta)}, \tag{A.6}
$$

see e.g. formula 6.2.2 in [4]. Therefore using formula (3.101) in [105] (here with $y := 0$, $\nu := 0$ and $\mu := s - \frac{1}{2}$, or see page 10 in [112]),

$$
\int_0^{+\infty} \frac{dt}{(a^2 + t^2)^{s+\frac{1}{2}}} = \frac{a^{-2s}}{2} B\left(\frac{1}{2}, s\right) {}_1F_2\left(\frac{1}{2}; 1 - s, \frac{1}{2}; 0\right)
$$

$$
= \frac{\Gamma\left(\frac{1}{2}\right) \Gamma(s)}{2a^{2s}\Gamma\left(\frac{1}{2} + s\right)}.
$$

Then, we recall that $\Gamma\left(\frac{1}{2}\right) = \pi^{\frac{1}{2}}$, so, making use of (A.5), for any small $a > 0$,

$$
\int_0^{+\infty} \frac{1 - \cos(2\pi t)}{(t^2 + a^2)^{s+\frac{1}{2}}} \, dt = \frac{\pi^{2s+\frac{1}{2}} \Gamma(1-s)}{2s\Gamma\left(s + \frac{1}{2}\right)} + \mathcal{O}(a^{2-2s}).
$$

Therefore, sending $a \to 0$ by the Dominated Convergence Theorem we obtain

$$
\int_0^{+\infty} \frac{1 - \cos(2\pi t)}{t^{1+2s}} \, dt = \lim_{a \to 0} \int_0^{+\infty} \frac{1 - \cos(2\pi t)}{(t^2 + a^2)^{s+\frac{1}{2}}} \, dt = \frac{\pi^{2s+\frac{1}{2}} \Gamma(1-s)}{2s\Gamma\left(s + \frac{1}{2}\right)}.
$$

$$\tag{A.7}$$

Now we recall the integral representation of the Beta function (see e.g. formulas 6.2.1 and 6.2.2 in [4]), namely

$$\frac{\Gamma\left(\frac{n-1}{2}\right)\Gamma\left(\frac{1}{2}+s\right)}{\Gamma\left(\frac{n}{2}+s\right)} = B\left(\frac{n-1}{2}, \frac{1}{2}+s\right) = \int_0^{+\infty} \frac{\tau^{\frac{n-3}{2}}}{(1+\tau)^{\frac{n}{2}+s}} \, d\tau. \qquad (A.8)$$

We also observe that in any dimension N the $(N-1)$-dimensional measure of the unit sphere is $\frac{N\pi^{\frac{N}{2}}}{\Gamma\left(\frac{N}{2}+1\right)}$, (see e.g. [89]). Therefore

$$\int_{\mathbb{R}^N} \frac{dY}{(1+|Y|^2)^{\frac{N+1+2s}{2}}} = \frac{N\pi^{\frac{N}{2}}}{\Gamma\left(\frac{N}{2}+1\right)} \int_0^{+\infty} \frac{\rho^{N-1}}{(1+\rho^2)^{\frac{N+1+2s}{2}}} \, d\rho.$$

In particular, taking $N := n-1$ and using the change of variable $\rho^2 =: \tau$,

$$\int_{\mathbb{R}^{n-1}} \frac{d\eta}{(1+|\eta|^2)^{\frac{n+2s}{2}}} = \frac{(n-1)\pi^{\frac{n-1}{2}}}{\Gamma\left(\frac{n-1}{2}+1\right)} \int_0^{+\infty} \frac{\rho^{n-2}}{(1+\rho^2)^{\frac{n+2s}{2}}} \, d\rho$$

$$= \frac{(n-1)\pi^{\frac{n-1}{2}}}{2\,\Gamma\left(\frac{n-1}{2}+1\right)} \int_0^{+\infty} \frac{\tau^{\frac{n-3}{2}}}{(1+\tau)^{\frac{n+2s}{2}}} \, d\tau.$$

By comparing this with (A.8), we conclude that

$$\int_{\mathbb{R}^{n-1}} \frac{d\eta}{(1+|\eta|^2)^{\frac{n+2s}{2}}} = \frac{(n-1)\pi^{\frac{n-1}{2}}}{2\,\Gamma\left(\frac{n-1}{2}+1\right)} \cdot \frac{\Gamma\left(\frac{n-1}{2}\right)\Gamma\left(\frac{1}{2}+s\right)}{\Gamma\left(\frac{n}{2}+s\right)}$$

$$= \frac{\pi^{\frac{n-1}{2}}\Gamma\left(\frac{1}{2}+s\right)}{\Gamma\left(\frac{n}{2}+s\right)}.$$

Accordingly, with the change of variable $\eta := |\omega_1|^{-1}(\omega_2, \ldots, \omega_n)$,

$$\int_{\mathbb{R}^n} \frac{1-\cos(2\pi\omega_1)}{|\omega|^{n+2s}} \, d\omega$$

$$= \int_{\mathbb{R}} \left(\int_{\mathbb{R}^{n-1}} \frac{1-\cos(2\pi\omega_1)}{(\omega_1^2 + \omega_2^2 + \cdots + \omega_n^2)^{\frac{n+2s}{2}}} \, d\omega_2 \ldots d\omega_n \right) d\omega_1$$

$$= \int_{\mathbb{R}} \left(\int_{\mathbb{R}^{n-1}} \frac{1-\cos(2\pi\omega_1)}{|\omega_1|^{1+2s}(1+|\eta|^2)^{\frac{n+2s}{2}}} \, d\eta \right) d\omega_1$$

$$= \frac{\pi^{\frac{n-1}{2}} \Gamma\left(\frac{1}{2}+s\right)}{\Gamma\left(\frac{n}{2}+s\right)} \int_{\mathbb{R}} \frac{1-\cos(2\pi\omega_1)}{|\omega_1|^{1+2s}} \, d\omega_1$$

$$= \frac{2\pi^{\frac{n-1}{2}} \Gamma\left(\frac{1}{2}+s\right)}{\Gamma\left(\frac{n}{2}+s\right)} \int_0^{+\infty} \frac{1-\cos(2\pi t)}{t^{1+2s}} \, dt.$$

Hence, recalling (A.7),

$$\int_{\mathbb{R}^n} \frac{1-\cos(2\pi\omega_1)}{|\omega|^{n+2s}} \, d\omega = \frac{2\pi^{\frac{n-1}{2}} \Gamma\left(\frac{1}{2}+s\right)}{\Gamma\left(\frac{n}{2}+s\right)} \cdot \frac{\pi^{2s+\frac{1}{2}} \Gamma(1-s)}{2s\Gamma\left(s+\frac{1}{2}\right)}$$

$$= \frac{\pi^{\frac{n}{2}+2s} \Gamma(1-s)}{s\Gamma\left(\frac{n}{2}+s\right)},$$

as desired.

To complete the picture, we also give a different proof of (A.7) which does not use the theory of special functions, but the Fourier transform in the sense of distributions and the unique analytic continuation:

Proof (Alternative proof of (A.7)) We recall (see[1] e.g. pages 156–157 in [141]) that for any $\lambda \in \mathbb{C}$ with $\Re\lambda \in (0, 1/2)$ we have

$$\mathscr{F}\left(|x|^{\lambda-1}\right) = \frac{\pi^{\frac{1}{2}-\lambda} \Gamma\left(\frac{\lambda}{2}\right)}{\Gamma\left(\frac{1-\lambda}{2}\right) |x|^{\lambda}}$$

in the sense of distribution, that is

$$\int_{\mathbb{R}} |x|^{\lambda-1}\hat{\phi}(x) \, dx = \frac{\pi^{\frac{1}{2}-\lambda} \Gamma\left(\frac{\lambda}{2}\right)}{\Gamma\left(\frac{1-\lambda}{2}\right)} \int_{\mathbb{R}} |x|^{-\lambda}\phi(x) \, dx$$

for every $\phi \in C_0^\infty(\mathbb{R})$. The same result is obtained in [22], in the proof of Proposition 2.4 b, pages 13–16.

As a consequence

$$\int_{\mathbb{R}} \left(\int_{\mathbb{R}} |x|^{\lambda-1}\phi(y) \, e^{-2\pi i xy} \, dy \right) dx = \frac{\pi^{\frac{1}{2}-\lambda} \Gamma\left(\frac{\lambda}{2}\right)}{\Gamma\left(\frac{1-\lambda}{2}\right)} \int_{\mathbb{R}} |x|^{-\lambda}\phi(x) \, dx. \qquad (A.9)$$

---

[1] To check (A.9) the reader should note that the normalization of the Fourier transform on page 138 in [141] is different than the one here in (2.1).

By changing variable $x \mapsto -x$ in the first integral, we also see that

$$\int_{\mathbb{R}} \left(\int_{\mathbb{R}} |x|^{\lambda-1} \phi(y) \, e^{2\pi i xy} \, dy \right) dx = \frac{\pi^{\frac{1}{2}-\lambda} \Gamma\left(\frac{\lambda}{2}\right)}{\Gamma\left(\frac{1-\lambda}{2}\right)} \int_{\mathbb{R}} |x|^{-\lambda} \phi(x) \, dx. \qquad (A.10)$$

By summing together (A.9) and (A.10), we obtain

$$\int_{\mathbb{R}} \left(\int_{\mathbb{R}} |x|^{\lambda-1} \phi(y) \, \cos(2\pi xy) \, dy \right) dx = \frac{\pi^{\frac{1}{2}-\lambda} \Gamma\left(\frac{\lambda}{2}\right)}{\Gamma\left(\frac{1-\lambda}{2}\right)} \int_{\mathbb{R}} |x|^{-\lambda} \phi(x) \, dx.$$

It is convenient to exchange the names of the integration variables in the first integral above: hence we write

$$\int_{\mathbb{R}} \left(\int_{\mathbb{R}} |y|^{\lambda-1} \phi(x) \, \cos(2\pi xy) \, dx \right) dy = \frac{\pi^{\frac{1}{2}-\lambda} \Gamma\left(\frac{\lambda}{2}\right)}{\Gamma\left(\frac{1-\lambda}{2}\right)} \int_{\mathbb{R}} |x|^{-\lambda} \phi(x) \, dx.$$

Now we fix $R > 0$, we point out that

$$\int_{\mathbb{R}} |y|^{\lambda-1} \chi_{(-R,R)}(y) \, dy = 2 \int_0^R y^{\lambda-1} \, dy = \frac{2R^\lambda}{\lambda},$$

and we obtain that

$$\int_{\mathbb{R}} \left(\int_{\mathbb{R}} |y|^{\lambda-1} \phi(x) \left(\cos(2\pi xy) - \chi_{(-R,R)}(y) \right) dx \right) dy$$
$$= \frac{\pi^{\frac{1}{2}-\lambda} \Gamma\left(\frac{\lambda}{2}\right)}{\Gamma\left(\frac{1-\lambda}{2}\right)} \int_{\mathbb{R}} |x|^{-\lambda} \phi(x) \, dx - \frac{2R^\lambda}{\lambda} \int_{\mathbb{R}} \phi(x) \, dx. \qquad (A.11)$$

This formula holds true, in principle, for $\lambda \in \mathbb{C}$, with $\Re\lambda \in (0, 1/2)$, but by the uniqueness of the analytic continuation in the variable λ it holds also for $\lambda \in (-2, 0)$.

Now we observe that the map

$$(x, y) \mapsto |y|^{\lambda-1} \phi(x) \left(\cos(2\pi xy) - \chi_{(-R,R)}(y) \right)$$

belongs to $L^1(\mathbb{R} \times \mathbb{R})$ if $\lambda \in (-2, 0)$, hence we can use Fubini's Theorem and exchange the order of the repeated integrals in (A.11): in this way, we deduce that

$$\int_{\mathbb{R}} \left(\int_{\mathbb{R}} |y|^{\lambda-1} \phi(x) \left(\cos(2\pi xy) - \chi_{(-R,R)}(y) \right) dy \right) dx$$
$$= \frac{\pi^{\frac{1}{2}-\lambda} \Gamma\left(\frac{\lambda}{2}\right)}{\Gamma\left(\frac{1-\lambda}{2}\right)} \int_{\mathbb{R}} |x|^{-\lambda} \phi(x) \, dx - \frac{2R^\lambda}{\lambda} \int_{\mathbb{R}} \phi(x) \, dx.$$

Since this is valid for every $\phi \in C_0^\infty(\mathbb{R})$, we conclude that

$$\int_\mathbb{R} |y|^{\lambda-1} \left(\cos(2\pi xy) - \chi_{(-R,R)}(y) \right) dy = \frac{\pi^{\frac{1}{2}-\lambda} \Gamma(\frac{\lambda}{2})}{\Gamma(\frac{1-\lambda}{2})} |x|^{-\lambda} - \frac{2R^\lambda}{\lambda}, \qquad \text{(A.12)}$$

for any $\lambda \in (-2,0)$ and $x \neq 0$. In this setting, we also have that

$$\int_\mathbb{R} |y|^{\lambda-1} \left(\chi_{(-R,R)}(y) - 1 \right) dy = -2 \int_R^{+\infty} |y|^{\lambda-1} dy = \frac{2R^\lambda}{\lambda}.$$

By summing this with (A.12), we obtain

$$\int_\mathbb{R} |y|^{\lambda-1} \left(\cos(2\pi xy) - 1 \right) dy = \frac{\pi^{\frac{1}{2}-\lambda} \Gamma(\frac{\lambda}{2})}{\Gamma(\frac{1-\lambda}{2})} |x|^{-\lambda}.$$

Hence, we take $\lambda := -2s \in (-2,0)$, and we obtain that

$$\int_\mathbb{R} \frac{\cos(2\pi xy) - 1}{|y|^{1+2s}} \, dy = \frac{\pi^{2s+\frac{1}{2}} \Gamma(-s)}{\Gamma(\frac{1}{2}+s)} |x|^{2s}.$$

By changing variable of integration $t := xy$, we obtain (A.7).

References

1. DDD: Discrete Dislocation Dynamics. https://www.ma.utexas.edu/mediawiki/index.php/Main_Page
2. Wikipedia page on Nonlocal equations. https://www.ma.utexas.edu/mediawiki/index.php/Main_Page
3. N. Abatangelo, E. Valdinoci, A notion of nonlocal curvature. Numer. Funct. Anal. Optim. **35**(7–9), 793–815 (2014)
4. M. Abramowitz, I.A. Stegun (eds.), *Handbook of Mathematical Functions with Formulas, Graphs, and Mathematical Tables*. A Wiley-Interscience Publication (Wiley, New York/National Bureau of Standards, Washington, 1984). Reprint of the 1972 edition, Selected Government Publications
5. L. Ambrosio, X. Cabré, Entire solutions of semilinear elliptic equations in $\mathbf{R}^3$ and a conjecture of De Giorgi. J. Am. Math. Soc. **13**(4), 725–739 (2000) (electronic)
6. L. Ambrosio, G. De Philippis, L. Martinazzi, Gamma-convergence of nonlocal perimeter functionals. Manuscr. Math. **134**(3–4), 377–403 (2011)
7. D. Applebaum, *Lévy Processes and Stochastic Calculus*, vol. 116 (Cambridge University Press, Cambridge/New York, 2009), pp. xxx+460
8. I. Athanasopoulos, L. Caffarelli, Continuity of the temperature in boundary heat control problems. Adv. Math. **224**(1), 293–315 (2010)
9. B. Barrios, A. Figalli, E. Valdinoci, Bootstrap regularity for integro-differential operators and its application to nonlocal minimal surfaces. Ann. Sc. Norm. Super. Pisa Cl. Sci.(5), **13**(3), 609–639 (2014)
10. B. Barrios, I. Peral, F. Soria, E. Valdinoci, A Widder's type theorem for the heat equation with nonlocal diffusion. Arch. Ration. Mech. Anal. **213**(2), 629–650 (2014)
11. R.F. Bass, M. Kassmann, Harnack inequalities for non-local operators of variable order. Trans. Am. Math. Soc. **357**(2), 837–850 (2005)
12. H. Berestycki, L. Caffarelli, L. Nirenberg, Further qualitative properties for elliptic equations in unbounded domains. Ann. Scuola Norm. Sup. Pisa Cl. Sci. (4) **25**(1–2), 69–94 (1998/1997). Dedicated to Ennio De Giorgi
13. J. Bertoin, *Lévy Processes*. Volume 121 of Cambridge Tracts in Mathematics (Cambridge University Press, Cambridge, 1996)
14. P. Biler, G. Karch, R. Monneau, Nonlinear diffusion of dislocation density and self-similar solutions. Commun. Math. Phys. **294**(1), 145–168 (2010)
15. C. Bjorland, L. Caffarelli, A. Figalli, Nonlocal tug-of-war and the infinity fractional Laplacian. Commun. Pure Appl. Math. **65**(3), 337–380 (2012)

16. P. Blanchard, E. Brüning, *Mathematical Methods in Physics*. Volume 26 of Progress in Mathematical Physics (Birkhäuser Boston, Boston, 2003), German edition. Distributions, Hilbert space operators, and variational methods

17. M. Bonforte, Y. Sire, J.L. Vázquez, Existence, uniqueness and asymptotic behavior for fractional porous medium equations on bounded domains (2014). arXiv preprint arXiv:1404.6195. To appear in Discr. Cont. Dyn. Sys. (2015)

18. M. Bonforte, J.L. Vázquez, Quantitative local and global a priori estimates for fractional nonlinear diffusion equations. Adv. Math. **250**, 242–284 (2014)

19. M. Bonforte, J.L. Vázquez, Fractional nonlinear degenerate diffusion equations on bounded domains part i. Existence, uniqueness and upper bounds (2015). arXiv preprint arXiv:1508.07871. To appear in Nonlinear Anal. TMA (2015)

20. M. Bonforte, J.L. Vázquez, A priori estimates for fractional nonlinear degenerate diffusion equations on bounded domains. Arch. Ration. Mech. Anal. **218**(1), 317–362 (2015)

21. J. Bourgain, H. Brezis, P. Mironescu, Another look at Sobolev spaces, in *Optimal Control and Partial Differential Equations*, ed. by J.L. Menaldi, E. Rofman, A. Sulem (IOS Press, Amsterdam, 2001), pp. 439–455. A volume in honor of A. Bensoussan's 60th birthday

22. C. Bucur, Some observations on the Green function for the ball in the fractional Laplace framework (2015). arXiv preprint arXiv:1502.06468

23. X. Cabré, E. Cinti, Energy estimates and 1-D symmetry for nonlinear equations involving the half-Laplacian. Discret. Contin. Dyn. Syst. **28**(3), 1179–1206 (2010)

24. X. Cabre, M.M. Fall, J. Solà-Morales, T. Weth, Curves and surfaces with constant nonlocal mean curvature: meeting Alexandrov and Delaunay (2015). arXiv preprint arXiv:1503.00469

25. X. Cabré, Y. Sire, Nonlinear equations for fractional Laplacians II: existence, uniqueness, and qualitative properties of solutions. Trans. Am. Math. Soc. **367**(2), 911–941 (2015)

26. X. Cabré, J. Solà-Morales, Layer solutions in a half-space for boundary reactions. Commun. Pure Appl. Math. **58**(12), 1678–1732 (2005)

27. L. Caffarelli, D. De Silva, O. Savin, No title yet (2015). arXiv preprint arXiv:1505.02304

28. L. Caffarelli, J.-M. Roquejoffre, O. Savin, Nonlocal minimal surfaces. Commun. Pure Appl. Math. **63**(9), 1111–1144 (2010)

29. L. Caffarelli, O. Savin, E. Valdinoci, Minimization of a fractional perimeter-Dirichlet integral functional. Ann. Inst. H. Poincaré Anal. Non Linéaire **32**(4), 901–924 (2015)

30. L. Caffarelli, L. Silvestre, An extension problem related to the fractional Laplacian. Commun. Partial Differ. Equ. **32**(7–9), 1245–1260 (2007)

31. L. Caffarelli, L. Silvestre, The Evans-Krylov theorem for nonlocal fully nonlinear equations. Ann. Math. (2) **174**(2), 1163–1187 (2011)

32. L. Caffarelli, P. Souganidis, Convergence of nonlocal threshold dynamics approximations to front propagation. Arch. Ration. Mech. Anal. **195**(1), 1–23 (2010)

33. L. Caffarelli, P. Stinga, Fractional elliptic equations: caccioppoli estimates and regularity (2014). arXiv preprint arXiv:1409.7721

34. L. Caffarelli, E. Valdinoci, Uniform estimates and limiting arguments for nonlocal minimal surfaces. Calc. Var. Partial Differ. Equ. **41**(1–2), 203–240 (2011)

35. L. Caffarelli, E. Valdinoci, Regularity properties of nonlocal minimal surfaces via limiting arguments. Adv. Math. **248**, 843–871 (2013)

36. L. Caffarelli, A. Vasseur, Drift diffusion equations with fractional diffusion and the quasi-geostrophic equation. Ann. Math. (2) **171**(3), 1903–1930 (2010)

37. L. Caffarelli, J.L. Vazquez, Nonlinear porous medium flow with fractional potential pressure. Arch. Ration. Mech. Anal. **202**(2), 537–565 (2011)

38. L.A. Caffarelli, J.L. Vázquez, Asymptotic behaviour of a porous medium equation with fractional diffusion. Discret. Contin. Dyn. Syst. **29**(4), 1393–1404 (2011)

39. M. Caputo, Linear models of dissipation whose Q is almost frequency independent. II. Fract. Calc. Appl. Anal. **11**(1), 4–14 (2008). Reprinted from Geophys. J. R. Astr. Soc. **13**(5), 529–539 (1967)

40. A. Chambolle, M. Morini, M. Ponsiglione, A nonlocal mean curvature flow and its semi-implicit time-discrete approximation. SIAM J. Math. Anal. **44**(6), 4048–4077 (2012)
41. Z.-Q. Chen, R. Song, Estimates on Green functions and Poisson kernels for symmetric stable processes. Math. Ann. **312**(3), 465–501 (1998)
42. E. Cinti, J. Serra, E. Valdinoci, Quantitative flatness results and BV-estimates for stable nonlocal minimal surfaces (2016). arXiv preprint arXiv:1602.00540
43. G. Ciraolo, A. Figalli, F. Maggi, M. Novaga, Rigidity and sharp stability estimates for hypersurfaces with constant and almost-constant nonlocal mean curvature (2015). arXiv preprint arXiv:1503.00653
44. M. Cozzi, E. Valdinoci, Plane-like minimizers for a non-local Ginzburg-Landau-type energy in a periodic medium (2015). arXiv preprint arXiv:1505.02304
45. J. Dávila, On an open question about functions of bounded variation. Calc. Var. Partial Differ. Equ. **15**(4), 519–527 (2002)
46. J. Dávila, M. del Pino, S. Dipierro, E. Valdinoci, Concentration phenomena for the nonlocal Schrödinger equation with Dirichlet datum (2015, to appear on Anal. PDE). arXiv preprint arXiv:1403.4435
47. J. Dávila, M. del Pino, J. Wei, Nonlocal minimal Lawson cones (2013). arXiv preprint arXiv:1303.0593
48. J. Dávila, M. del Pino, J. Wei, Nonlocal s-minimal surfaces and Lawson cones (2014). arXiv preprint arXiv:1402.4173
49. Dávila, M. del Pino, S. Dipierro, E. Valdinoci, Nonlocal Delaunay surfaces. Nonlinear Anal.: Theory Methods Appl. (2015)
50. E. De Giorgi, Sulla convergenza di alcune successioni d'integrali del tipo dell'area. Rend. Mat. (6) **8**, 277–294 (1975). Collection of articles dedicated to Mauro Picone on the occasion of his ninetieth birthday
51. R. de la Llave, E. Valdinoci, A generalization of Aubry-Mather theory to partial differential equations and pseudo-differential equations. Ann. Inst. H. Poincaré Anal. Non Linéaire **26**(4), 1309–1344 (2009)
52. R. de la Llave, E. Valdinoci, Symmetry for a Dirichlet-Neumann problem arising in water waves. Math. Res. Lett. **16**(5), 909–918 (2009)
53. A. de Pablo, F. Quirós, A. Rodríguez, J.L. Vázquez, A fractional porous medium equation. Adv. Math. **226**(2), 1378–1409 (2011)
54. M. del Pino, M. Kowalczyk, J. Wei, A counterexample to a conjecture by De Giorgi in large dimensions. C. R. Math. Acad. Sci. Paris **346**(23–24), 1261–1266 (2008)
55. A. Di Castro, T. Kuusi, G. Palatucci, Nonlocal Harnack inequalities. J. Funct. Anal. **267**(6), 1807–1836 (2014)
56. A. Di Castro, B. Ruffini, N. Matteo, E. Valdinoci, Non-local isoperimetric problems (2015). arXiv preprint arXiv:1406.7545. To appear on Calc. Var. Partial Differential Equations
57. E. Di Nezza, G. Palatucci, E. Valdinoci, Hitchhiker's guide to the fractional Sobolev spaces. Bull. Sci. Math. **136**(5), 521–573 (2012)
58. S. Dipierro, A. Figalli, G. Palatucci, E. Valdinoci, Asymptotics of the s-perimeter as $s \searrow 0$. Discret. Contin. Dyn. Syst. **33**(7), 2777–2790 (2013)
59. S. Dipierro, A. Figalli, E. Valdinoci, Strongly nonlocal dislocation dynamics in crystals. Commun. Partial Differ. Equ. **39**(12), 2351–2387 (2014)
60. S. Dipierro, G. Palatucci, E. Valdinoci, Dislocation dynamics in crystals: a macroscopic theory in a fractional Laplace setting. Commun. Math. Phys. **333**(2), 1061–1105 (2015)
61. S. Dipierro, O. Savin, E. Valdinoci, A nonlocal free boundary problem (2014). arXiv preprint arXiv:1411.7971. Accepted by SIAM J. Math. Anal.
62. S. Dipierro, O. Savin, E. Valdinoci, All functions are locally s-harmonic up to a small error (2015). arXiv preprint arXiv:1404.3652. To appear on J. Eur. Math. Soc. (JEMS)
63. S. Dipierro, O. Savin, E. Valdinoci, Boundary behavior of nonlocal minimal surfaces (2015). arXiv preprint arXiv:1506.04282
64. S. Dipierro, O. Savin, E. Valdinoci, Graph properties for nonlocal minimal surfaces (2015). arXiv preprint arXiv:1506.04281

65. S. Dipierro, E. Valdinoci, Continuity and density results for a one-phase nonlocal free boundary problem (2015). arXiv preprint arXiv:1504.05569

66. G. Duvaut, J.-L. Lions, Neravenstva v mekhanike i fizike (1980), p. 384. Translated from the French by S. Yu. Prishchepionok and T.N. Rozhkovskaya

67. B. Dyda, Fractional Hardy inequality with a remainder term. Colloq. Math. **122**(1), 59–67 (2011)

68. B. Dyda, Fractional calculus for power functions and eigenvalues of the fractional Laplacian. Fract. Calc. Appl. Anal. **15**(4), 536–555 (2012)

69. L.C. Evans, *Partial Differential Equations.* Volume 19 of Graduate Studies in Mathematics, 2nd edn. (American Mathematical Society, Providence, 2010)

70. M.M. Fall, V. Felli, Unique continuation property and local asymptotics of solutions to fractional elliptic equations. Commun. Partial Differ. Equ. **39**(2), 354–397 (2014)

71. M.M. Fall, S. Jarohs, Overdetermined problems with fractional Laplacian. ESAIM: COCV **21**(4), 924–938 (2015). doi:10.1051/cocv/2014048. http://dx.doi.org/10.1051/cocv/2014048

72. A. Farina, Symmetry for solutions of semilinear elliptic equations in $\mathbf{R}^N$ and related conjectures. Atti Accad. Naz. Lincei Cl. Sci. Fis. Mat. Natur. Rend. Lincei (9) Mat. Appl. **10**(4), 255–265 (1999)

73. A. Farina, E. Valdinoci, The state of the art for a conjecture of De Giorgi and related problems, in *Recent Progress on Reaction-Diffusion Systems and Viscosity Solutions* (World Scientific Publishing, Hackensack, 2009), pp. 74–96

74. A. Farina, E. Valdinoci, Rigidity results for elliptic PDEs with uniform limits: an abstract framework with applications. Indiana Univ. Math. J. **60**(1), 121–141 (2011)

75. M. Felsinger, Parabolic equations associated with symmetric nonlocal operators. Ph.D. thesis Dissertation, 2013

76. M. Felsinger, M. Kassmann, Local regularity for parabolic nonlocal operators. Commun. Partial Differ. Equ. **38**(9), 1539–1573 (2013)

77. F. Ferrari, B. Franchi, Harnack inequality for fractional sub-Laplacians in Carnot groups. Math. Z. **279**(1–2), 435–458 (2015)

78. A. Figalli, N. Fusco, F. Maggi, V. Millot, M. Morini, Isoperimetry and stability properties of balls with respect to nonlocal energies. Commun. Math. Phys. **336**(1), 441–507 (2015)

79. A. Figalli, E. Valdinoci, Regularity and Bernstein-type results for nonlocal minimal surfaces (2013). arXiv preprint arXiv:1307.0234. To appear on J. Reine Angew. Math.

80. N. Forcadel, C. Imbert, R. Monneau, Homogenization of some particle systems with two-body interactions and of the dislocation dynamics. Discret. Contin. Dyn. Syst. **23**(3), 785–826 (2009)

81. R.L. Frank, M. del Mar González, D.D. Monticelli, J. Tan, An extension problem for the CR fractional Laplacian. Adv. Math. **270**, 97–137 (2015)

82. R.L. Frank, E. Lenzmann, Uniqueness of non-linear ground states for fractional Laplacians in $\mathbb{R}$. Acta Math. **210**(2), 261–318 (2013)

83. R.L. Frank, E. Lenzmann, L. Silvestre, Uniqueness of radial solutions for the fractional laplacian (2013). arXiv preprint arXiv:1302.2652

84. A. Friedman, PDE problems arising in mathematical biology. Netw. Heterog. Media **7**(4), 691–703 (2012)

85. N. Ghoussoub, C. Gui, On a conjecture of De Giorgi and some related problems. Math. Ann. **311**(3), 481–491 (1998)

86. E. Giusti, *Minimal Surfaces and Functions of Bounded Variation.* Volume 80 of Monographs in Mathematics (Birkhäuser Verlag, Basel, 1984)

87. M. del Mar González, R. Monneau, Slow motion of particle systems as a limit of a reaction-diffusion equation with half-Laplacian in dimension one. Discret. Contin. Dyn. Syst. **32**(4), 1255–1286 (2012)

88. G. Grubb, Fractional Laplacians on domains, a development of Hörmander's theory of μ-transmission pseudodifferential operators. Adv. Math. **268**, 478–528 (2015)

89. G. Huber, Notes: gamma function derivation of n-sphere volumes. Am. Math. Mon. **89**(5), 301–302 (1982)

90. N.E. Humphries, N. Queiroz, J.R.M. Dyer, N.G. Pade, M.K. Musyl, K.M. Schaefer, D.W. Fuller, J.M. Brunnschweiler, J.D.R. Doyle, K. Thomas, J.D. Houghton, et al. Environmental context explains Lévy and Brownian movement patterns of marine predators. Nature **465**(7301), 1066–1069 (2010)
91. C. Imbert, Level set approach for fractional mean curvature flows. Interfaces Free Bound. **11**(1), 153–176 (2009)
92. S. Jarohs, Symmetry via maximum principles for nonlocal nonlinear boundary value problems. Ph.D. thesis Dissertation, 2015
93. C.-Y. Kao, Y. Lou, W. Shen, Random dispersal vs. non-local dispersal. Discret. Contin. Dyn. Syst. **26**(2), 551–596 (2010)
94. M. Kassmann, The classical Harnack inequality fails for non-local operators. Sonderforschungsbereich 611 (2007, preprint)
95. A. Kiselev, F. Nazarov, A. Volberg, Global well-posedness for the critical 2D dissipative quasi-geostrophic equation. Invent. Math. **167**(3), 445–453 (2007)
96. A. Koldobsky, *Fourier Analysis in Convex Geometry*. Volume 116 of Mathematical Surveys and Monographs (American Mathematical Society, Providence, 2005)
97. T. Kulczycki, Properties of Green function of symmetric stable processes. Probab. Math. Statist. **17**(2, Acta Univ. Wratislav. No. 2029), 339–364 (1997)
98. T. Kuusi, G. Mingione, Y. Sire, A fractional Gehring lemma, with applications to nonlocal equations. Atti Accad. Naz. Lincei Rend. Lincei Mat. Appl. **25**(4), 345–358 (2014)
99. T. Kuusi, G. Mingione, Y. Sire, Nonlocal equations with measure data. Commun. Math. Phys. **337**(3), 1317–1368 (2015)
100. T. Kuusi, G. Mingione, Y. Sire, Nonlocal self-improving properties. Anal. PDE **8**(1), 57–114 (2015)
101. N. Laskin, Fractional quantum mechanics and Lévy path integrals. Phys. Lett. A **268**(4–6), 298–305 (2000)
102. C. Limmaneevichitr, Dislocation motion is analogous to the movement of caterpillar (2009). Available from Youtube, https://youtu.be/08a9hNFj22Y
103. L. Lombardini, Fractional perimeter and nonlocal minimal surfaces (2015). arXiv preprint arXiv:1508.06241
104. A. Massaccesi, E. Valdinoci, Is a nonlocal diffusion strategy convenient for biological populations in competition? (2015). arXiv preprint arXiv:1503.01629
105. R.J. Mathar, Yet another table of integrals (2012). arXiv preprint arXiv:1207.5845
106. V. Maz'ya, Lectures on isoperimetric and isocapacitary inequalities in the theory of Sobolev spaces, in *Heat Kernels and Analysis on Manifolds, Graphs, and Metric Spaces (Paris, 2002)*. Volume 338 of Contemporary Mathematics (American Mathematical Society, Providence, 2003), pp. 307–340
107. R. Metzler, J. Klafter, The random walk's guide to anomalous diffusion: a fractional dynamics approach. Phys. Rep. **339**(1), 77 (2000)
108. S.A. Molčanov, E. Ostrovskiĭ, Symmetric stable processes as traces of degenerate diffusion processes. Teor. Verojatnost. i Primenen. **14**, 127–130 (1969)
109. G.M. Bisci, V.D. Radulescu, R. Servadei, *Variational Methods for Nonlocal Fractional Problems*, vol. 162 (Cambridge University Press, Cambridge, 2016)
110. E. Montefusco, B. Pellacci, G. Verzini, Fractional diffusion with Neumann boundary conditions: the logistic equation. Discret. Contin. Dyn. Syst. Ser. B **18**(8), 2175–2202 (2013)
111. R. Musina, A.I. Nazarov, On fractional Laplacians. Commun. Partial Differ. Equ. **39**(9), 1780–1790 (2014)
112. F. Oberhettinger, *Tabellen zur Fourier Transformation* (Springer, Berlin/Göttingen/ Heidelberg, 1957)
113. A. de Pablo, F. Quirós, A. Rodríguez, J.L. Vázquez, A general fractional porous medium equation. Commun. Pure Appl. Math. **65**(9), 1242–1284 (2012)
114. G. Palatucci, O. Savin, E. Valdinoci, Local and global minimizers for a variational energy involving a fractional norm. Ann. Mat. Pura Appl. (4) **192**(4), 673–718 (2013)

115. S. Patrizi, E. Valdinoci, Crystal dislocations with different orientations and collisions. Arch. Ration. Mech. Anal. **217**(1), 231–261 (2015)
116. S. Patrizi, E. Valdinoci, Relaxation times for atom dislocations in crystals (2015). arXiv preprint arXiv:1504.00044
117. A.C. Ponce, *Elliptic PDEs, Measures and Capacities. From the Poisson equation to Nonlinear Thomas-Fermi problems*, vol. 23 (European Mathematical Society (EMS), Zürich, 2015)
118. A. Reynolds, C. Rhodes, The Lévy flight paradigm: random search patterns and mechanisms. Ecology **90**(4), 877–887 (2009)
119. X. Ros-Oton, Nonlocal elliptic equations in bounded domains: a survey (2015). arXiv preprint arXiv:1504.04099
120. O. Savin, Regularity of flat level sets in phase transitions. Ann. Math. (2) **169**(1), 41–78 (2009)
121. O. Savin, Minimal surfaces and minimizers of the Ginzburg-Landau energy. Contemp. Math. **528**, 43–57 (2010)
122. O. Savin, E. Valdinoci, Density estimates for a nonlocal variational model via the Sobolev inequality. SIAM J. Math. Anal. **43**(6), 2675–2687 (2011)
123. O. Savin, E. Valdinoci, Γ-convergence for nonlocal phase transitions. Ann. Inst. H. Poincaré Anal. Non Linéaire **29**(4), 479–500 (2012)
124. O. Savin, E. Valdinoci, Regularity of nonlocal minimal cones in dimension 2. Calc. Var. Partial Differ. Equ. **48**(1–2), 33–39 (2013)
125. O. Savin, E. Valdinoci, Density estimates for a variational model driven by the Gagliardo norm. J. Math. Pures Appl. (9) **101**(1), 1–26 (2014)
126. W. Schoutens, *Lévy Processes in Finance: Pricing Financial Derivatives* (Wiley, Chichester/West Sussex/New York, 2003)
127. R.W. Schwab, L. Silvestre, Regularity for parabolic integro-differential equations with very irregular kernels (2014). arXiv preprint arXiv:1412.3790
128. R. Servadei, E. Valdinoci, On the spectrum of two different fractional operators. Proc. R. Soc. Edinb. Sect. A **144**(4), 831–855 (2014)
129. R. Shambayati, Z. Zieleźny, On Fourier transforms of distributions with one-sided bounded support. Proc. Am. Math. Soc. **88**(2), 237–243 (1983)
130. L. Silvestre, Hölder estimates for solutions of integro-differential equations like the fractional Laplace. Indiana Univ. Math. J. **55**(3), 1155–1174 (2006)
131. L. Silvestre, Regularity of the obstacle problem for a fractional power of the Laplace operator. Commun. Pure Appl. Math. **60**(1), 67–112 (2007)
132. Y. Sire, E. Valdinoci, Fractional Laplacian phase transitions and boundary reactions: a geometric inequality and a symmetry result. J. Funct. Anal. **256**(6), 1842–1864 (2009)
133. N. Soave, E. Valdinoci, Overdetermined problems for the fractional Laplacian in exterior and annular sets (2014). arXiv preprint arXiv:1412.5074
134. P.R. Stinga, J.L. Torrea, Extension problem and Harnack's inequality for some fractional operators. Commun. Partial Differ. Equ. **35**(11), 2092–2122 (2010)
135. E. Valdinoci, A fractional framework for perimeters and phase transitions. Milan J. Math. **81**(1), 1–23 (2013)
136. J. Van Schaftingen, M. Willem, *Set Transformations, Symmetrizations and Isoperimetric Inequalities* (Springer, Milan, 2004), pp. 135–152
137. J.L. Vázquez, Nonlinear diffusion with fractional Laplacian operators, in *Nonlinear Partial Differential Equations*. Volume 7 of Abel Symposium (Springer, Heidelberg, 2012), pp. 271–298
138. J.L. Vázquez, Recent progress in the theory of nonlinear diffusion with fractional Laplacian operators. Discret. Contin. Dyn. Syst. Ser. S **7**(4), 857–885 (2014)
139. A. Visintin, Generalized coarea formula and fractal sets. Jpn. J. Ind. Appl. Math. **8**(2), 175–201 (1991)
140. G.M. Viswanathan, V. Afanasyev, S.V. Buldyrev, E.J. Murphy, P.A. Prince, H.E. Stanley, Lévy flight search patterns of wandering albatrosses. Nature **381**(6581), 413–415 (1996)

141. W.F. Donoghue Jr., *Distributions and Fourier Transforms*. Pure and Applied Mathematics, vol. 32 (Academic, New York/London, 1969), 315 p.
142. W.A. Woyczyński, Lévy processes in the physical sciences, in *Lévy Processes* (Birkhäuser Boston, Boston, 2001), pp. 241–266
143. K. Yosida, *Functional Analysis*. Classics in Mathematics (Springer, Berlin, 1995). Reprint of the 6th edn. (1980)

LECTURE NOTES OF THE UNIONE MATEMATICA ITALIANA

Editor in Chief: Ciro Ciliberto and Susanna Terracini

Editorial Policy

1. The UMI Lecture Notes aim to report new developments in all areas of mathematics and their applications - quickly, informally and at a high level. Mathematical texts analysing new developments in modelling and numerical simulation are also welcome.

2. Manuscripts should be submitted to
   Redazione Lecture Notes U.M.I.
   umi@dm.unibo.it
   and possibly to one of the editors of the Board informing, in this case, the Redazione about the submission. In general, manuscripts will be sent out to external referees for evaluation. If a decision cannot yet be reached on the basis of the first 2 reports, further referees may be contacted. The author will be informed of this. A final decision to publish can be made only on the basis of the complete manuscript, however a refereeing process leading to a preliminary decision can be based on a pre-final or incomplete manuscript. The strict minimum amount of material that will be considered should include a detailed outline describing the planned contents of each chapter, a bibliography and several sample chapters.

3. Manuscripts should in general be submitted in English. Final manuscripts should contain at least 100 pages of mathematical text and should always include

   - a table of contents;
   - an informative introduction, with adequate motivation and perhaps some historical remarks: it should be accessible to a
     reader not intimately familiar with the topic treated;
   - a subject index: as a rule this is genuinely helpful for the reader.

4. For evaluation purposes, please submit manuscripts in electronic form, preferably as pdf- or zipped ps-files. Authors are asked, if their manuscript is accepted for publication, to use the LaTeX2e style files available from Springer's web-server at
   ftp://ftp.springer.de/pub/tex/latex/svmonot1/ for monographs
   and at
   ftp://ftp.springer.de/pub/tex/latex/svmultt1/ for multi-authored volumes

5. Authors receive a total of 50 free copies of their volume, but no royalties. They are entitled to a discount of 33.3% on the price of Springer books purchased for their personal use, if ordering directly from Springer.

6. Commitment to publish is made by letter of intent rather than by signing a formal contract. Springer-Verlag secures the copyright for each volume. Authors are free to reuse material contained in their LNM volumes in later publications: A brief written (or e-mail) request for formal permission is sufficient.

Printed in the United States
By Bookmasters